Lecture Notes in Statistics

Vol. 1: R. A. Fisher: An Appreciation. Edited by S. E. Fienberg and D. V. Hinkley. XI, 208 pages, 1980.

Vol. 2: Mathematical Statistics and Probability Theory. Proceedings 1978. Edited by W. Klonecki, A. Kozek, and J. Rosiński. XXIV, 373 pages, 1980.

Vol. 3: B. D. Spencer, Benefit-Cost Analysis of Data Used to Allocate Funds. VIII, 296 pages, 1980.

Vol. 4: E. A. van Doorn, Stochastic Monotonicity and Queueing Applications of Birth-Death Processes. VI, 118 pages, 1981.

Vol. 5: T. Rolski, Stationary Random Processes Associated with Point Processes. VI, 139 pages, 1981.

Vol. 6: S. S. Gupta and D.-Y. Huang, Multiple Statistical Decision Theory: Recent Developments. VIII, 104 pages, 1981.

Vol. 7: M. Akahira and K. Takeuchi, Asymptotic Efficiency of Statistical Estimators. VIII, 242 pages, 1981.

Vol. 8: The First Pannonian Symposium on Mathematical Statistics. Edited by P. Révész, L. Schmetterer, and V. M. Zolotarev. VI, 308 pages, 1981.

Vol. 9: B. Jørgensen, Statistical Properties of the Generalized Inverse Gaussian Distribution. VI, 188 pages, 1981.

Vol. 10: A. A. McIntosh, Fitting Linear Models: An Application on Conjugate Gradient Algorithms. VI, 200 pages, 1982.

Vol. 11: D. F. Nicholls and B. G. Quinn, Random Coefficient Autoregressive Models: An Introduction. V, 154 pages, 1982.

Vol. 12: M. Jacobsen, Statistical Analysis of Counting Processes. VII, 226 pages, 1982.

Vol. 13: J. Pfanzagl (with the assistance of W. Wefelmeyer), Contributions to a General Asymptotic Statistical Theory. VII, 315 pages, 1982.

Vol. 14: GLIM 82: Proceedings of the International Conference on Generalised Linear Models. Edited by R. Gilchrist. V, 188 pages, 1982.

Vol. 15: K. R. W. Brewer and M. Hanif, Sampling with Unequal Probabilities. IX, 164 pages, 1983.

Vol. 16: Specifying Statistical Models: From Parametric to Non-Parametric, Using Bayesian or Non-Bayesian Approaches. Edited by J. P. Florens, M. Mouchart, J. P. Raoult, L. Simar, and A. F. M. Smith. XI, 204 pages, 1983.

Vol. 17: I. V. Basawa and D. J. Scott, Asymptotic Optimal Inference for Non-Ergodic Models. IX, 170 pages, 1983.

Vol. 18: W. Britton, Conjugate Duality and the Exponential Fourier Spectrum. V, 226 pages, 1983.

Vol. 19: L. Fernholz, von Mises Calculus For Statistical Functionals. VIII, 124 pages, 1983.

Vol. 20: Mathematical Learning Models – Theory and Algorithms: Proceedings of a Conference. Edited by U. Herkenrath, D. Kalin, W. Vogel. XIV, 226 pages, 1983.

Vol. 21: H. Tong, Threshold Models in Non-linear Time Series Analysis. X, 323 pages, 1983.

Vol. 22: S. Johansen, Functional Relations, Random Coefficients and Nonlinear Regression with Application to Kinetic Data. VIII, 126 pages. 1984.

Vol. 23: D. G. Saphire, Estimation of Victimization Prevalence Using Data from the National Crime Survey. V, 165 pages. 1984.

Vol. 24: T. S. Rao, M. M. Gabr, An Introduction to Bispectral Analysis and Bilinear Time Series Models. VIII, 280 pages, 1984.

Vol. 25: Time Series Analysis of Irregularly Observed Data. Proceedings, 1983. Edited by E. Parzen. VII, 363 pages, 1984.

ctd. on inside back cover

Lecture Notes in Statistics

Edited by D. Brillinger, S. Fienberg, J. Gani, J. Hartigan, and K. Krickeberg

41

François Baccelli
Pierre Brémaud

Palm Probabilities and Stationary Queues

Springer-Verlag

Authors

François Baccelli
INRIA
B.P. 105, Domaine de Voluceau-Rocquencourt
78153 Le Chesnay Cedex, France

Pierre Brémaud
Laboratoire des Signaux et Systèmes
CNRS/E.S.E., Plateau du Moulon
91190 Gif-sur-Yvette, France

AMS Subject Classification (1980): 60K25, 60G55

ISBN 978-0-387-96514-7 ISBN 978-1-4615-7561-0 (eBook)
DOI 10.1007/978-1-4615-7561-0

2147/3140-543210

INTRODUCTION

In order to obtain explicit results in Queueing Theory, it is sometimes necessary to make specific assumptions which render the mathematical model analytically tractable. For this reason, Markovian, renewal, or semi-Markov queueing models are most frequently used and of great practical value. However, there are some results which are true in general, only requiring stationarity of the model. This is the case of Little's formula $Q = \lambda W$ and of many others.

This monograph is mainly a tutorial intended to provide an introduction to the theoretical framework in which such formulae are obtained. It was originally inspired by the book "Queues and Point Processes" by Franken, König, Arndt and Schmidt and by Neveu's approach to the existence problem of stationary states (see references and comments).

The first part of the monograph is devoted to a detailed review of the basic results concerning Palm probabilities and useful to Queueing Theory.

In most texts on Point Process Theory, the treatment of the so-called Palm theory is usually not presented in the first chapters. In our opinion it is an elementary and intuitive theory directly accessible to a reader without background in Point Process Theory, and we have devoted a third of the monograph to it. We have also included a few items which are not directly useful to Queueing Theory, such as Feller's paradox (§ 4.2) and the connection between Papangelou's and Watanabe's approach to conditional (stochastic) intensity (I, §7).

As a consequence many results of Queueing Theory given in Part II, find a very simple proof and can be considered as application exercises of the theory of Part I.

Part II features 3 topics : the queueing formulae of all kinds, the existence of stationary states, and the insensitivity theory.

Our treatment of insensitivity is rather exhaustive from the theoretical point of view. It contains in particular an extension of the Palm theory of insensitivity to the case with infinite state space and the converse theorem of insensitivity. However we have not harvested all the known results of queueing theory on the subject, such as for instance the insensitivity of networks (BCMP, Kelly, etc...). The bibliography will help the reader to broaden his culture in this direction. The same can be said of our treatment of queueing formulae.

Concerning the construction of stationary states, the theory has only recently been impulsed by the East German school around Franken and König and the French school around Neveu. Although the above two approaches are similar in many respects, we have chosen to present Neveu's approach in terms of the "θ_t-formalism". Our treatment in the present monograph is, by necessity, rather incomplete since the theory has obviously more to say and is presently being developed. We have chosen a few typical problems which can serve as an introduction to this area of research. Here again the bibliography will guide the reader for further research.

Acknowledgements

This monograph owes much to J. Neveu, D. Flipo, M. Klein and Ph. Robert who presented portions of its material in a joint INRIA - Paris VI workshop (1982-84) organized by G. Fayolle and the authors.

Daniel Flipo, Jim Pitman, Jean Walrand and Marc Yor reviewed portions of the manuscript, providing helpful comments and necessary corrections.

Florence Deshors successfully typed this monograph from a rather difficult handwritten manuscript.

The authors are grateful to all the above mentioned people, as well as to Springer-Verlag for its efficient handling of the manuscript.

TABLES OF CONTENTS

PART I

STATIONARY POINT PROCESSES

AND

PALM PROBABILITIES

1 - STATIONARY MARKED POINT PROCESS

1.1 The canonical space of a point process on $\mathbb{R}$

A counting measure on $\mathbb{R}$ is a measure m on $(\mathbb{R}, \mathbf{B})$ such that

(i) $m(C) \in \{0,1,\ldots,\infty\}$ for all $C \in B$,

(ii) $m([a,b]) < \infty$ for all $a,b \in \mathbb{R}$.

Let M be the set of all counting measures on $\mathbb{R}$. The σ-field on M generated by the functions $m \to m(C)$, $C \in B$ is denoted by $\mathbf{M}$. $(M,\mathbf{M})$ is the canonical space of point processes on $\mathbb{R}$.

Associated with each counting measure $m \in M$, is a unique sequence $(t_n, n \in \mathbb{Z})$ of $\overline{\mathbb{R}}$ such that

$$m(.) = \sum_{n\in\mathbb{Z}} \delta_{t_n}(.)$$

$$-\infty \leq \ldots \leq t_{-1} \leq t_0 \leq 0 < t_1 \leq t_2 \leq \ldots \leq +\infty$$

where δ_x is the Dirac measure at $x \in \mathbb{R}$, with the convention $\delta_{\infty}(.) = \delta_{-\infty}(.) \equiv 0$.

The time t_n is called the nth point of m. The mapping $m \to t_n$ is a random variable on $(M,\mathbf{M})$. The counting measure is called simple iff $m(\{x\}) = 0$ or 1 for all $x \in \mathbb{R}$. Clearly m is simple if and only if $t_n < t_{n+1}$ for all $n \in \mathbb{Z}$ such that $|t_n| < \infty$.

For any function $f : \mathbb{R} \to \mathbb{R}_+$

$$\int_{\mathbb{R}} f(x)m(dx) = \sum_{n\in\mathbb{Z}} f(t_n)\, 1_{\mathbb{R}}(t_n).$$

The translation operator T_t operating on M is defined for each $t \in \mathbb{R}$ by

$$T_t m(C) = m(C+t),\ C \in B.$$

The family $(T_t, t \in \mathbb{R})$ forms a flow of $(M,\mathbf{M})$ i.e.

(i) $(t,m) \to T_t m$ is measurable with respect to $B \times \mathbf{M}$ and $\mathbf{M}$,

(ii) $T_t \circ T_s = T_{t+s}$ for all $t, s \in \mathbb{R}$. In particular $T_0 = I$ (identity) and $T_t^{-1} = T_{-t}$.

1.2 Stationary point processes

Let $(\Omega, \mathbf{F})$ be a measurable space and N a measurable mapping from $(\Omega, \mathbf{F})$ into $(M,\mathbf{M})$. N is called a point process.

Thus, for each $\omega \in \Omega$, $N(\omega)$ is a counting measure. Define $N(C)$ by $N(\omega,C)=N(\omega)(C)$ and T_n by $T_n(\omega) = t_n(N(\omega))$.

If a probability measure P is given on $(\Omega, \mathbf{F})$, the pair (N, P) is called a <u>stochastic point process.</u>

Let $(\theta_t, t \in \mathbb{R})$ be a flow on $(\Omega, \mathbf{F})$, i.e.

(i) $(t,\omega) \to \theta_t(\omega)$ is measurable with respect to $B \times F$ and $\mathbf{F}$,

(ii) θ_t is bijective for all $t \in \mathbb{R}$,

(iii) $\theta_t \circ \theta_s = \theta_{t+s}$ for all $t, s \in \mathbb{R}$. In particular θ_0 = identity and $\theta_t^{-1}=\theta_{-t}$.

If (N, P) is a stochastic point process such that for all $t \in \mathbb{R}$

$$\begin{cases} N(\theta_t(\omega)) = T_t N(\omega) \quad (\omega \in \Omega) \\ P \circ \theta_t = P, \end{cases}$$

the triple (N, θ_t, P) is called a <u>stationary stochastic point process</u>, or stationary point process for short.

<u>Example 1 : Stationary point process in canonical form.</u> Here $(\Omega, \mathbf{F}) = (M, \mathbf{M})$, $\theta_t = T_t$ $(t \in \mathbb{R})$, and $P = \mathbf{P}$ where P is a measure on $(M, \mathbf{M})$ such that $P \circ T_t = \mathbf{P}$ $(t \in \mathbb{R})$. Taking N = identity, we obtain a stationary point process which is said to be in canonical form, namely $(\Phi, T_t, \mathbf{P})$ if Φ denotes the identity map of M. □

<u>Example 2 : Superposition of independent point processes in canonical form.</u> Let $(M_i, \mathbf{M}_i, T_t^{(i)}, \Phi_i)$ $(1 \le i \le k)$ be k replicas of $(M, \mathbf{M}, T_t, \Phi)$ and for each i $(1 \le i \le k)$ let P_i be a probability on $(M_i, \mathbf{M}_i)$ which is $T_t^{(i)}$-invariant for all $t \in \mathbb{R}$.

Define the product space

$$(\Omega, \mathbf{F}, P) = \left(\prod_{i=1}^{k} M_i,\ \bigotimes_{i=1}^{k} \mathbf{M}_i,\ \bigotimes_{i=1}^{k} P_i \right)$$

and, for each $t \in \mathbb{R}$:

$$\theta_t = \bigotimes_{i=1}^{k} T_t^{(i)}.$$

The above notation means $\theta_t \omega = (T_t^{(i)} m_i \ ;\ 1 \le i \le k)$ where $\omega = (m_i \ ;\ 1 \le i \le k)$.

Define

$$N(\omega) = \sum_{i=1}^{k} m_i,\ N_i(\omega) = m_i \ (1 \le i \le k).$$

The stationary point process (N,θ_t,P) is called the superposition of the stationary point processes (N_i,θ_t,P) $(1 \leq i \leq k)$. □

1.3 Introduction of the marks

Let $Z = (Z_n, n \in \mathbb{Z})$ be a sequence of measurable mappings from $(\Omega,\mathcal{F})$ into some measurable space $(K,\mathcal{K})$. It is called a sequence of marks of (N,θ) iff

$$Z_n(\omega) = Z_0(\theta_{T_n}\omega).$$

If moreover (N,θ_t,P) is a stationary point process, $((N,Z),\theta_\cdot,P)$ is called a stationary marked point process (with marks in K).

Example 1 : Let (N,θ_t,P) be a stationary point process. Define for each $n \in \mathbb{Z}$, each $\omega \in \Omega$,

$$\tilde{Z}_n(\omega) = \theta_{T_n(\omega)}(\omega) \tag{1.3.2}$$

or, more briefly, $\tilde{Z}_n = \theta_{T_n}$. Then $\tilde{Z} = (\tilde{Z}_n, n \in \mathbb{Z})$ is called the universal sequence of marks of (N,θ_t) since every sequence of marks $Z = (Z_n, n \in \mathbb{Z})$ is of the form $Z_n = Z_0(\theta_{T_n}(\omega)) = Z_0(\tilde{Z}_n(\omega))$ for all $n \in \mathbb{N}$, all $\omega \in \Omega$. □

Example 2 : Let (N,θ_t,P) be a stationary point process and let $(A_t, t \in \mathbb{R})$ be a non-negative process on $(\Omega,\mathcal{F})$ such that

$$A_t = A_0(\theta_t),\ t \in \mathbb{R}. \tag{1.3.3}$$

The formula

$$Z_n = \int_{T_n}^{T_{n+1}} A_s\,ds \tag{1.3.4}$$

defines a sequence of marks of (N,θ_t,P). For instance if $A_t = 1$, $Z_n = T_{n+1}-T_n$. □

Example 3 : Let E be a denumerable state space. Let $\Omega = D(\mathbb{R};E)$ be the set of functions $\omega : \mathbb{R} \to E$ which are right-continuous, and with a σ-discrete set of discontinuities. Let $\theta_t : \Omega \to \Omega$ be defined by

$$(\theta_t(\omega))(u) = \omega(u+t). \tag{1.3.5}$$

Let F be the σ-field generated by the coordinate mappings $(X_t, t \in \mathbb{R})$ (i.e. $X_t(\omega)=\omega(t)$).

The discontinuities of $(X_t, t \in \mathbb{R})$ define a point process N by $N(B) = \sum_{s \in B} 1(X_s \neq X_{s-})$ and clearly $N\circ\theta_t = T_tN$ $(t \in \mathbb{R})$. Also if T_n is the nth point of N, $Z_n = X_{T_n}$ defines a sequence of marks $Z = (Z_n, n \in \mathbb{Z})$.

Let P be a probability on $(\Omega, \mathbf{F})$ such that $P\circ\theta_t = P$ $(t \in \mathbb{R})$. Then $((N,Z),\theta_\cdot,P)$ is a stationary marked point process.

Suppose that under P, $(X_t, t \in \mathbb{R})$ is a (regular) Markov chain with infinitesimal generator $\mathbf{Q} = \{q_{ij} ; i,j \in E\}$ and stationary distribution $\pi = (\pi(i) ; i \in E)$. Recall that $\pi'\mathbf{Q} = 0$, where prime denotes transposition.

Let H be a subset of E x E - diag(E x E). Let N_H be the point process counting the H-transitions of $(X_t ; t \in \mathbb{R})$, i.e.

$$N_H(C) = \int_C 1_H(X_{s-},X_s)N(ds) = \sum_{s\in C} 1_H(X_{s-},X_s). \qquad (1.3.6)$$

(N_H, θ_t, P) is a stationary point process. In particular for H = E x E - diag(E x E), we have $N_H = N$ the "basic" point process of the chain, which counts all transitions.

Let $(T_n, n \in \mathbb{Z})$ be the jump times of $(X_t, t \in \mathbb{R})$, i.e. the points of the basic point process N. The formula

$$Z_n = X_{T_n} \quad (n \in \mathbb{Z}) \qquad (1.3.7)$$

defines a sequence of marks for (N, θ_t) with values in the state space.

Clearly (N, Z) and $(X_t, t \in \mathbb{R})$ are equivalent descriptions of the same object. ◻

<u>Example 4</u> : <u>Canonical space of marked point processes</u>

It is somewhat easier to understand the θ_t-formalism in terms of canonical spaces :

Let $(K,\mathbf{K})$ be a measurable space and define $\tilde{M}_k$ to be the set of sequences :

$$((t_n,k_n), n \in \mathbb{Z}), \qquad (1.3.8)$$

where $t_n \in \overline{\mathbb{R}}$, $k_n \in K$ $(n \in \mathbb{Z})$, and

$$-\infty \le \dots \le t_{-2} \le t_{-1} \le t_0 \le 0 < t_1 \le t_2 \le \dots \le +\infty. \qquad (1.3.9)$$

Define a measure μ on $(\mathbb{R} \times K, \mathbf{B} \times \mathbf{K})$ by

$$\mu(C\times L) = \sum_{n\in\mathbb{Z}} 1_C(t_n)\, 1_L(k_n) \quad (C \in \mathbf{B},\ L \in \mathbf{K}). \qquad (1.3.10)$$

Call M_k the set of measures μ representable by (1.3.8) to (1.3.10). Let $\mathbf{M}_k$ be the σ-field on M_k generated by $(\mu(C\times L) : C \in \mathbf{B},\ L \in \mathbf{K})$. For each $t \in \mathbb{R}$, the operator T_t acts on a measure μ on $(\mathbb{R} \times K; B \times \mathbf{K})$ in the following manner :

$$T_t\,\mu(C\times L) = \mu((C-t)\times L), \quad (C \in B,\ L \in \mathbf{K}). \qquad (1.3.11)$$

$(T_t,\ t \in \mathbb{R})$ is a flow on $(M_k, \mathbf{M}_k)$ and $\mu \to t_n$, $\mu \to k_n$ are $\mathbf{M}_k$-measurable mappings.*

Let P be a probability on $(M_k, \mathbf{M}_k)$ such that

$$P \circ T_t = \mathbf{P} \quad (t \in \mathbb{R}). \tag{1.3.12}$$

Call ψ the identity mapping on M_k : $\psi(\mu) = \mu$ and T_n, Z_n the mappings $\mu \to t_n$, $\mu \to k_n$ respectively.

Then with the identifications

$$\begin{cases} (\Omega, \mathbf{F}, P) \equiv (M_k, \mathbf{M}_k, \mathbf{P}) \\ \theta_t = T_t \\ N(.) = \psi(. \times K), \end{cases} \tag{1.3.13}$$

we have a stationary point process (N, θ_t, P). Moreover $Z = (Z_n,\ n \in \mathbb{Z})$ is a sequence of marks associated with (N, θ_t). □

1.4 Two properties of stationary point processes

(1.4.1) Independent stationary point processes have no common point

Consider the situation described in Example 1.2.2 with k = 2. It will be shown that N_1 and N_2 cannot have common points. In other words, two independent stationary point processes have almost surely no common point. The proof is given in the case where one of them (say N_1) has a finite intensity, i.e. $E[N_1(0,1]] < \infty$. Then $E[N_1\{t\})] = 0$ $(t \in \mathbb{R})$, by using stationarity.

Now using the above remark and Fubini's theorem, we get

$$E[\int_{\mathbb{R}} N_1(\{s\})\ N_2(ds)] = \int_{M_1}\int_{M_2}\int_{\mathbb{R}} m_1(\{s\})\ m_2(ds)\ \mathbf{P}_1(dm_1)\ \mathbf{P}_2(dm_2)$$

$$= \int_{M_2}\int_{\mathbb{R}} [\int_{M_1} m_1(\{s\})\ \mathbf{P}_1(dm_1)]\ m_2(ds)\mathbf{P}_2(dm_2) = 0.$$

Therefore $\int_{\mathbb{R}} N_1(\{s\})\ N_2(ds) = 0$, P.a.s., which proves the assertion. □

---

* All proofs concerning measurability are omitted. The reader may consult the classical references on point processes for this (see bibliography).

(1.4.2) Let (N, θ_t, P) be a stationary point process. Then

$$P(\{N(\mathbb{R}) = 0\} \cup \{N(0,\infty) = N(-\infty,0) = +\infty\}) = 1.$$

Indeed, let $H_s = \{N(s,\infty) = 0\}$. Since $\theta_t H_s = H_{s-t}$ for all $s,t \in \mathbb{R}$, $P(H_t) =$ constant. Also H_t increases with t, so that $\bigcup_{n=0}^{\infty} H_n \supset \bigcap_{n=1}^{\infty} H_{-n}$. But $G = \bigcup_{n=1}^{\infty} H_n = \{N(0,\infty) < \infty\}$, $H = \bigcap_{n=0}^{\infty} H_{-n} = \{N(\mathbb{R}) = 0\}$. Therefore $P(G-H)=0$. A similar reasoning based on $\tilde{H}_s = \{N(-\infty,s) = 0\}$ leads to $P(G'-H) = 0$, where $G'=\{N(-\infty,0) < \infty\}$. Therefore $P((G \cup G')-H) = 0$, q.e.d. ∎

2 INTENSITY

2.1 Intensity of a stationary point process

Let (N, θ_t, P) be a stationary point process. The non-negative (possibly infinite) number

$$\lambda = E[N((0,1])] \tag{2.1.1}$$

is called the intensity of (N, θ_t, P). More generally, define for all $C \in B$

$$\mu(C) = E[N(C)]. \tag{2.1.2}$$

From the stationarity of (N,θ_t,P), $\mu(C+t) = \mu(C)$ and therefore μ defines a translation invariant measure on $(\mathbb{R}, \mathbf{B})$. It is therefore proportional to the Lebesgue measure ℓ so that

$$\mu(C) = \lambda . \ell(C). \tag{2.1.3}$$

ASSUMPTIONS

> In this monograph we will assume that the stationary point processes considered have a finite intensity and moreover that this intensity is not null (if it is null, all points T_n are at infinity). Moreover it will be assumed that the point processes herein are simple, i.e. $P(N\{T_n\})>1)=0$ for all $n \in \mathbb{Z}$, and that $P(N(\mathbb{R})=\infty)=1$.

These assumptions will generally not be recalled in the text.

2.2 Intensity measure of a stationary marked point process

Let $((N,Z), \theta_t, P)$ be a stationary marked point process with marks in $(K,\mathbf{K})$, defined on $(\Omega, \mathbf{F})$.

Define the random kernel N_Z from $(\Omega, \mathbf{F})$ into $(\mathbb{R} \times K, \mathbf{B} \times \mathbf{K})$ by

$$N_Z(C \times L) = \int_C 1_L(Z_0 \circ \theta_s)\, N(ds) := \sum_{n \in \mathbb{Z}} 1_L(Z_n)\, 1_C(T_n) \tag{2.2.1}$$

where $C \in B$, $L \in \mathbf{K}$. Observe that, by construction

$$N_Z(\theta_t(\omega), C \times L) = N_Z(\omega, (C+t) \times L).$$

The intensity measure λ_Z associated with the stationary marked point process $((N, Z), \theta_t, P)$ is a measure on $(\mathbb{R} \times K, \mathbf{B} \times \mathbf{K})$ defined by

$$\lambda_Z(C \times L) = E[N_Z(C \times L)] \tag{2.2.2}$$

where $C \in B$, $L \in \mathbf{K}$.

If the intensity λ of (N, θ_t, P) is finite, which we assume throughout the present monograph, λ_Z is a σ-finite measure. Indeed since $N_Z(C \times L) \leq N(C)$, $\lambda_Z(C \times L) \leq \lambda \ell(C)$ for all $C \in B$, all $L \in \mathbf{K}$.

From the θ_t-stationarity of P

$$\lambda_Z((C+t) \times L) = \lambda_Z(C \times L).$$

Thus, for fixed $L \in \mathbf{K}$, $\lambda_Z(\cdot \times L)$ is a shift invariant σ-finite measure on $\mathbb{R}$, and must therefore be proportional to the Lebesgue measure. Necessarily

$$\lambda_Z(C \times L) = \lambda_Z((0,1] \times L)\ \ell(C).$$

Let μ_Z be the measure on $(K, \mathbf{K})$ defined by

$$\mu_Z(L) = \frac{1}{\lambda} \lambda_Z((0,1] \times L). \tag{2.2.3}$$

Then, summarizing the above results and definitions

$$\lambda_Z(C \times L) = \lambda . \ell(C) . \mu_Z(L). \tag{2.2.4}$$

$\mu_Z(L)$ is the average number of points T_n in $C \in B$ such that $Z_n \in L$, divided by the average number of points T_n in C. This characterization does not depend upon $C \in B$ such that $\ell(C) > 0$.

μ_Z is a probability measure, called for reasons to appear soon, the Palm probability distribution of the marks Z_n.

3 PALM PROBABILITIES

3.1 - Definition

Let $\tilde{Z}$ be the sequence of universal marks (§ 1.3) associated with (N, θ_t, P) and denote by $\tilde{\lambda}$ the intensity measure $\lambda_{\tilde{Z}}$ associated with $((N, \tilde{Z}), \theta_t, P)$. Also denote $\mu_{\tilde{Z}}$ by P_N^0.

Thus, recalling that the marks $\tilde{Z}_n = \theta_{T_n}$ take their values in $(\Omega, \mathbf{F})$, we get

$$P_N^0(A) = \frac{1}{\lambda\, \ell(C)} E[\int_C 1_A \circ \theta_s . N(ds)] = \frac{1}{\lambda \ell(C)} E[\sum_{n \in \mathbb{Z}} 1_A \circ \theta_{T_n} 1_C(T_n)]. \qquad (3.1.1)$$

P_N^0 is a probability measure on $(\Omega, \mathbf{F})$ called the <u>Palm probability</u> of the stationary point process (N, θ_t, P) with <u>finite intensity</u> λ.

One immediately observes that

$$P_N^0(T_0 = 0) = 1. \qquad (3.1.2)$$

This equality follows by taking $A = \{T_0 = 0\}$ in (3.1.1) and observing that $\int_C 1_{\{T_0 = 0\}} \circ \theta_s . N(ds) = \sum_{n \in \mathbb{Z}} 1(T_0 \circ \theta_{T_n} = 0) 1_C(T_n) = \sum_{n \in \mathbb{Z}} 1_C(T_n)$ since $T_0 \circ \theta_{T_n} \equiv 0$.

Let now Z be an arbitrary sequence of marks associated with (N, θ_t, P) with values in $(K, \mathbf{K})$, and let $L \in \mathbf{K}$. Taking $A = \{Z_0 \in L\}$ in (3.1.1) and observing that $1_{\{Z_0 \in L\}} \circ \theta_{T_n} = 1(Z_0 \circ \theta_{T_n} \in L) = 1(Z_n \in L)$, we see that

$$\mu_z(L) = P_N^0(Z_0 \in L). \qquad (3.1.3)$$

Thus formula (3.1.1) takes the form :

$$\lambda_z(C \times L) = \lambda . \ell(C)\, P_N^0(Z_0 \in L). \qquad (3.1.4)$$

<u>Heuristic remarks</u>

In order to give some intuition as to what is going on, take (N, θ_t, P) to be the canonical stationary point process $(\Phi, T_t, \mathbf{P})$ of example 1, §1.2. Definition (3.1.1) reads

$$\mathbf{P}_\Phi^0(A) = \frac{1}{\lambda \ell(C)} E_P[\sum_{n \in \mathbb{Z}} 1_A \circ T_{t_n} 1_C(t_n)], \qquad (3.1.5)$$

where $A \in \mathbf{M}$.

A set $A \in \mathbf{M}$ can be viewed as a property which a given trajectory $m \in M$ does or does not satisfy. $1_A(T_{t_n} m) = 1$ when the trajectory m, viewed by an observer placed in t_n, satisfies A.

It will be seen later (in §8) that if $(\mathbf{P}, T_t)$ is ergodic (and using $\lambda\, \ell(C) = E_P[\Phi(C)]$)

$$\frac{1}{\lambda \ell(C)} E_P[\sum_{n \in \mathbb{Z}} 1_A \circ T_{t_n} 1_C(t_n)] = \lim_{N \to \infty} \frac{1}{N} \sum_{k=1}^{N} 1_A(T_{t_k} m) \;\; P.a.s. \qquad (3.1.6)$$

Therefore $P_\Phi^0(A)$ is in this case the limiting empiral frequency, under P, that an observer jumping from point to point, sees property A realized.

The first historical interpretation of Palm properties (to be given in §6.2) says that

$$P_\Phi^0(A) = \lim_{h \downarrow 0} \mathbf{P}\,(m_{T_{t_1}} \text{ verifies } A \mid t_1 \leq h). \tag{3.1.7}$$

This is called the local interpretation of Palm probability.

3.2 Invariance of the Palm probability

The Palm probability P_N^0 of the stationary point process (N, θ_t, P) has all its mass concentrated on $\Omega_0 = \{T_0 = 0\}$. On the other hand, on Ω_0, $\theta_{T_n} = (\theta_{T_1})^n$ for all $n \in \mathbb{Z}$. Therefore defining

$$\theta = \theta_{T_1}, \tag{3.2.1}$$

we see that θ is a bijection of Ω_0 and that

$$\theta_{T_n} = \theta^n \text{ on } \Omega_0, \text{ for all } n \in \mathbb{Z}. \tag{3.2.2}$$

In addition :

$$P_N^0 \text{ is } \theta\text{-invariant}. \tag{3.2.3}$$

Remark : This is expected in view of the heuristic interpretation in the previous subsection (see (3.1.6)).

Proof : Take $A \in \mathbf{F}$, $A \subset \Omega_0$

$$| P_N^0(A) - P_N^0(\theta^{-1}(A)) | \leq \frac{1}{\lambda t} E \,| \sum_{n \in \mathbb{Z}} (1_A \circ \theta^n - 1_{\theta^{-1}A} \circ \theta^n) 1_{(0,t]}(T_n) | ,$$

where we have applied the defining formula (3.1.1) to $C = (0,t]$.

Since $1_{\theta^{-1}A} \circ \theta^n = 1_A \circ \theta^{n+1}$, we see that

$$| P_N^0(A) - P_N^0(\theta^{-1}(A)) | \leq \frac{2}{\lambda t} .$$

Now let $t \to \infty$ to obtain

$$P_N^0(A) = P_N^0(\theta^{-1}A),$$

for all $A \in \mathbf{F}$, $A \subset \Omega_0$.

3.3 Campbell's formula

Let v be the real valued function defined on $\mathbb{R} \times \Omega$ by

$$v(\omega,t) = 1_A(\omega)\, 1_C(t),$$

where $A \in \mathcal{F}$, $C \in \mathcal{B}$. By definition of the product measure $P_N^0(d\omega)\times dt$ and of the measure $P(d\omega)\, N(\omega,dt)$, the defining equality (3.1.1) reads

$$\lambda \int_{\Omega\times\mathbb{R}} v(\omega,t)\, P_N^0(d\omega)dt = \int_{\Omega\times\mathbb{R}} v(\theta_t(\omega),t)\, P(d\omega)\, N(\omega,dt). \tag{3.3.1}$$

By standard arguments, (3.3.1) remains true for all non-negative measurable functions v from $(\Omega \times \mathbb{R},\ \mathcal{F} \times \mathcal{B})$ into $(\mathbb{R},\mathcal{B})$.

This is known as the generalized Campbell's formula.

The original Campbell's formula is obtained by specializing (3.3.1) to

$$v(\omega,t) = f(t,Z_0(\omega)),$$

where Z is a sequence of marks associated with (N,θ_t,P) with values in $(K,\mathcal{K})$.

In view of (3.1.4) :

$$\lambda \int_{\Omega\times\mathbb{R}} f(t,Z_0(\omega))\, P_N^0(d\omega)dt = \int_{\mathbb{R}\times K} f(t,Z)\, \lambda_Z(dt\times dz).$$

On the other hand

$$\int_{\Omega\times\mathbb{R}} f(t,Z_0(\theta_t(\omega)))\, P(d\omega)\, N(\omega,dt) = E[\sum_{n\in\mathbb{Z}} f(T_n,Z_0(\theta_{T_n}))]$$

$$= E[\sum_{n\in\mathbb{Z}} f(T_n,Z_n)].$$

Therefore (Campbell's formula)

$$E[\sum_{n\in\mathbb{Z}} f(T_n,Z_n)] = \int_{\mathbb{R}\times K} f(t,z)\, \lambda_Z(dt\times dz), \tag{3.3.2}$$

for all non-negative measurable functions f from $(\mathbb{R}\times K,\ \mathcal{B}\times\mathcal{K})$ into $(\mathbb{R},\mathcal{B})$.

3.4 The cycle formula and Wald's identity

Let (N,θ_t,P) and (N',θ_t,P) be two stationary point processes with finite intensities λ and λ' respectively. Note that N and N' are "jointly" stationary, in the sense that their stationarity is relative to the same $(\Omega,\mathcal{F},P,\theta_t)$.

The following formula, called the exchange formula, holds :

$$\lambda\ E_N^0[f] = \lambda'\ E_{N'}^0\left[\int_{[0,T_1')} f\circ\theta_t\ N(dt)\right], \tag{3.4.1}$$

for all non-negative measurable function $f : (\Omega, \mathbf{F}) \to (\mathbb{R}, \mathbf{B})$. Here T_n' is the nth point of N'.

For a proof see Neveu [2].

By the $\theta_{T_n'}$ stationarity of $P_{N'}^0$ the quantity

$$E_{N'}^0[f\circ\theta_{T_n'}\ N(\{T_n'\})]$$

does not depend upon $n \in \mathbb{Z}$, so that in (3.4.1), the integral can be taken over $(0, T_1']$ instead of $[0, T_1')$.

A special case of (3.4.1) is obtained for $f \equiv 1$, yielding

$$\lambda'\ E_{N'}^0[N((0,T_1'])] = \lambda \tag{3.4.2}$$

and therefore the exchange formula takes the much expected form

$$E_N^0[f] = \frac{E_{N'}^0\left[\int_{(0,T_1']} f\circ\theta_t\ N(dt)\right]}{E_{N'}^0\ [N((0,T_1'])]} \tag{3.4.3}$$

<u>Remark 1</u> : Formula (3.4.3) is rather intuitive in the ergodic context, if we can write

$$\begin{cases} E_N^0[f] = \lim_{n\uparrow\infty} \frac{1}{n} \sum_{k=1}^{n} f\circ\theta_{T_k} & \text{P.a.s.} \\ E_{N'}^0[g] = \lim_{n\uparrow\infty} \frac{1}{n} \sum_{k=1}^{n} g\circ\theta_{T_k} & \text{P.a.s.} \end{cases} \tag{3.4.4}$$

with $g = \int_{(T_0', T_1']} f\circ\theta_t \; N(dt)$ or $g = N((T_0', T_1'])$. See §8 for a discussion of ergodicity and the validity of (3.4.4). As a matter of fact, the "proof" of (3.4.3) by (3.4.4) is a familiar exercise for queueing theorists in the context, for instance, of busy cycles. See Part 2, §3.4. This is why we also call the exchange formula the <u>cycle formula</u>. □

<u>Remark 2</u> : Formula (3.4.3) can also be viewed as a generalization of <u>Wald's identity</u> since it reads

$$E_{N'}^0 \left[\sum_{k=0}^{N(0,T_1']} f\circ\theta_{T_k} \right] = E_{N'}^0 [N(0,T_1']] \, E_N^0[f]. \tag{3.4.5}$$

4 FROM PALM PROBABILITY TO STATIONARY PROBABILITY

4.1 The inversion formulae

Let h be a non-negative measurable function from $(\Omega\times\mathbb{R}, \mathcal{F}\times\mathcal{B})$ into $(\mathbb{R},\mathcal{B})$ such that for all $\omega \in \Omega$

$$\int_{\mathbb{R}} h(\omega,t)\, N(\omega,dt) = 1$$

and let f be a non-negative random variable on $(\Omega, \mathcal{F})$.

The generalized Campbell's formula (3.3.1) applied to $v(\omega,t) = h(\theta_{-t}\omega,t)\, f(\theta_{-t}\omega)$ yields

$$E[f] = \lambda \int_{\Omega\times\mathbb{R}} h(\theta_{-t}\omega,t)\, f(\theta_{-t}\omega)\, P_N^0(d\omega)dt. \tag{4.1.1}$$

If for instance we take $h(\omega,t) = 1_{[T_0(\omega),0)}$, we obtain the first inversion formula

$$E[f] = \lambda E_N^0 \left[\int_0^{T_1} f\circ\theta_t \; dt \right], \tag{4.1.2a}$$

which in the special case $f = 1_A$ $(A \in \mathcal{F})$ reads

$$P(A) = \lambda \int_0^{\infty} P_N^0(T_1 > t,\ \theta_t \in A)dt. \tag{4.1.2b}$$

The next formulae are obvious consequences of the θ_{T_1}-invariance of P_N^0 and of (4.1.2a) and (4.1.2b).

We obtain the second inversion formula

$$E[f] = \lambda\, E_N^0 \left[\int_0^{-T_{-1}} f \circ \theta_{-t}\, dt \right] \qquad (4.1.3a)$$

or, for $f = 1_A$ $(A \in F)$,

$$P(A) = \lambda \int_0^{\infty} P_N^0(-T_{-1} > t,\ \theta_{-t} \in A)\,dt. \qquad (4.1.3b)$$

More generally,

$$E[f] = \lambda\, E_N^0 \left[\int_{T_n}^{T_{n+1}} f \circ \theta_t\, dt \right] \qquad (4.1.3c)$$

and

$$P(A) = \lambda \int_0^{\infty} P_N^0(T_n < t \le T_{n+1},\ \theta_t \in A)\,dt. \qquad (4.1.3d)$$

4.2 Feller's paradox

Taking $f = 1$ in (4.1.2a) yields

$$\lambda\, E_N^0 [T_1] = 1. \qquad (4.2.1)$$

Let F_0 be the cumulative distribution function of T_1, under P_N^0

$$F_0(x) = P_N^0 (T_1 \le x). \qquad (4.2.2)$$

Letting $A = \{T_1 > v,\ -T_0 > w\}$ in (4.1.2b), with $v,w \in \mathbb{R}_+$, we obtain

$$P(T_1 > v,\ -T_0 > w) = \lambda \int_{v+w}^{\infty} (1-F_0(u))du. \qquad (4.2.3)$$

In particular, with $v=0$, and observing that, by construction of T_1, $P(T_1 > 0) = 1$

$$P(-T_0 > w) = \lambda \int_w^{\infty} (1-F_0(u))du \qquad (4.2.4a)$$

This shows in particular that $P(-T_0 > 0) = 1$ since $\lambda \int_0^{\infty} (1-F_0(u))du = \lambda E_N^0[T_1] = 1$. Similarly, taking $w = 0$ in (4.2.3), we get

$$P(T_1 > v) = \lambda \int_v^{\infty} (1-F_0(u))du. \qquad (4.2.4b)$$

Thus $-T_0$ and T_1 are identically distributed under P.

The case of renewal processes

Let (N, θ_t, P) be a stationary point process with finite intensity λ and P_N^0 be its Palm probability. Suppose moreover that under P_N^0 the inter-event sequence $(S_n, n \in \mathbb{Z})$ defined by

$$S_n = T_{n+1} - T_n \quad (n \in \mathbb{Z}) \tag{4.2.5}$$

is i.i.d. Then (N, P_N^0) is called a renewal process and (N, P) a delayed renewal process.

The existence of such a mathematical object is granted by the results of the forthcoming §4.4.

(4.2.6) The distribution of the sequence $S^* = (S_n, n \in \mathbb{Z} - \{0\})$ is the same under P and P_N^0.

Proof Let $g : (\mathbb{R}^{\mathbb{Z}}, \mathbf{B}^{\mathbb{Z}}) \to (\mathbb{R}, \mathbf{B})$ be an arbitrary non negative measurable function. It suffices to show that

$$E[g(S^*)] = E_N^0[g(S^*)].$$

By the inversion formula :

$$E[g(S^*)] = \lambda\, E_N^0\left[\int_0^{T_1} g(S^*(\theta_u))du\right].$$

But if $u \in [0, T_1)$, $S(\theta_u) = S$, so that

$$E[g(S^*)] = \lambda\, E_N^0\left[\int_0^{T_1} g(S^*)du\right] = \lambda\, E_N^0[T_1 g(S^*)] = \lambda\, E_N^0[T_1]\, E_N^0[g(S^*)] = E_N^0[g(S^*)],$$

where we have used the independence of $T_1 = S_0$ and S^* under P_N^0.

Next we shall show that

(4.2.7) S_0 and S^* are P-independent.

Proof : Computations similar to the above yield

$$E[f(S_0)\, g(S^*)] = \lambda\, E_N^0[f(S_0)T_1\, g(S^*)]$$

$$= \lambda\, E_N^0[f(S_0)T_1]\, E_N^0[g(S^*)] = \lambda E_N^0[f(S_0)T_1]E[g(S^*)] = E[f(S_0)]E[g(S^*)]. \qquad \square$$

So far the following was proven : under P, $S = (S_n, n \in \mathbb{Z})$ is a sequence of independent random variables, and $S^* = (S_n, n \in \mathbb{Z} - \{0\})$ is an i.i.d. sequence. Can we expect that $S = (S_n, n \in \mathbb{Z})$ is i.i.d. under P, i.e. that S_0 has the same distribution under P and under P_N^0. The answer is negative in general : from

$$E[f(S_0)] = \lambda\, E_N^0[f(S_0)T_1] = \lambda\, E_N^0[f(S_0)S_0],$$

we obtain (by taking $f(S_0) = S_0$, and recalling that $\lambda = 1/E_N^0[S_0]$)

$$E[S_0] = \frac{1}{E_N^0[S_0]}\, E_N^0[S_0^2].$$

Therefore, if $E[S_0] = E_N^0[S_0]$,

$$E_N^0[S_0]^2 = E_N^0[S_0^2].$$

The variance of S_0 has hence to be zero under P_N^0, i.e. S_n = constant, P_N^0-a.s. Indeed, one can check from (4.2.3) that in this particular case, S_0 has the same deterministic distribution under P and P_N^0.

4.3 The mean value formulae

Let (N, θ_t, P) be a stationary point process with finite intensity, and let P_N^0 be the associated Palm probability.

Let $(Z_t,\ t \in \mathbb{R})$ be a stochastic process with values in a measurable space $(E, \mathbf{E})$ and such that

$$Z_t = Z_0 \circ \theta_t, \quad (t \in \mathbb{R}). \tag{4.3.1}$$

Then, for any non-negative measurable function $g : (E, \mathbf{E}) \to (\mathbb{R}, \mathbf{B})$

$$E[g(Z_0)] = \frac{E_N^0[\int_0^{T_1} g(Z_t)dt]}{E_N^0[T_1]} \tag{4.3.2}$$

and

$$E_N^0[g(Z_0)] = \frac{E[\sum_{n \in \mathbb{Z}} g(Z_n) 1_{\{T_n \in (0,1]\}}]}{E[\sum_{n \in \mathbb{Z}} 1_{\{T_n \in (0,1]\}}]}. \tag{4.3.3}$$

These formulae just rephrase the inversion formula (4.1.2a) and the definition formula (3.1.1) of P_N^0.

4.4 The inverse construction

Define on $(\Omega, \mathbf{F})$ a flow $(\theta_t,\ t \in \mathbb{R})$ and a point process N such that $N(\theta_t) = T_t N$.

Let P^0 be a probability measure on $(\Omega, \mathbf{F})$ such that

$$P^0(\Omega_0) = 1, \tag{4.4.1}$$

where $\Omega_0 = \{T_0 = 0\}$, and

$$P^0(\theta_{T_n} \in .) = P^0(.), \quad (n \in \mathbb{Z}). \tag{4.4.2}$$

Moreover we will assume that

$$0 < E^0[T_1] < \infty. \tag{4.4.3}$$

We will see that P^0 is the Palm probability P_N^0 associated with the stationary point process (N,θ_t,P) for some probability P that is θ_t-invariant for all $t \in \mathbb{R}$. Moreover, in view of the inversion formulae, P will be unique.

As the inversion formula requires, if such a P exists, it should satisfy

$$P(A) = \frac{1}{E^0[T_1]} E^0[\int_0^{T_1} 1_A o\theta_t \, dt]. \tag{4.4.4}$$

Clearly (4.4.4) defines a probability on $(\Omega,\mathcal{F})$. We must show that P is θ_t-invariant for all $t \in \mathbb{R}$, and that

$$P_N^0 = P^0. \tag{4.4.5}$$

We first remark that, on Ω_0,

$$\int_{T_0 o\theta_{T_j}}^{T_1 o\theta_{T_j}} (1_A o\theta_s o\theta^j)ds = \int_{T_j}^{T_{j+1}} 1_A o\theta_s \, ds, \quad (j\in\mathbb{Z})$$

and therefore, since P^0 is θ_{T_n}-stationary

$$P(A) = \frac{1}{n}\frac{1}{E^0[T_1]} E^0[\int_0^{T_n} 1_A o\theta_s \, ds].$$

Also, for any $t \in \mathbb{R}$

$$P(\theta_t A) = \frac{1}{n}\frac{1}{E^0[T_1]} E^0[\int_0^{T_n} 1_{\theta_t A} o\theta_s \, ds] = \frac{1}{n}\frac{1}{E^0[T_1]} E^0[\int_{-t}^{T_n-t} 1_A o\theta_s \, ds].$$

Therefore

$$| P(A)-P(\theta_t A) | \le \frac{1}{n}\frac{1}{E^0[T_1]} E^0 | \int_{T_n-t}^{T_n} 1_A o\theta_s \, ds - \int_{-t}^{0} 1_A o\theta_s ds | \le \frac{2t}{E^0[T_1]}\frac{1}{n}.$$

Since this is true for all n, $P(A) = P(\theta_t A)$.

From (4.4.4)

$$E[f] = \frac{1}{E^0[T_1]} E^0\left[\int_0^{T_1} f\circ\theta_t \, dt\right].$$

In particular,

$$\frac{E[N((0,\varepsilon])]}{\varepsilon} = \frac{1}{E^0[T_1]} \frac{1}{\varepsilon} E^0\left[\int_0^{T_1} N((t,t+\varepsilon])dt\right] = \frac{1}{E^0[T_1]} \frac{1}{\varepsilon} E^0\left[\int_{T_1-\varepsilon}^{T_1} N(t,t+\varepsilon]dt\right] -$$

and therefore $\lim_{\varepsilon\downarrow 0} \frac{E[N(0,\varepsilon])]}{\varepsilon} = \frac{1}{E^0[T_1]}$. Since $\frac{E[N((0,\varepsilon])]}{\varepsilon} = \lambda$, we get

$$\lambda = \frac{1}{E^0[T_1]},$$

and therefore, in view of assumption (4.4.3), $0 < \lambda < \infty$. We can hence define the Palm probability P_N^0 associated with (N, θ_t, P). By the inversion formula

$$P(\theta_{T_0} \in A) = \lambda\, E_N^0\left[\int_0^{T_1} (1_A \circ\theta_{T_0+t})dt\right]$$

$$= \lambda\, E_N^0\left[\int_0^{T_1} 1_A \, dt\right] = \lambda\, E_N^0[T_1 \, 1_A],$$

where we have observed that $1_A\circ\theta_{T_0+t} = 1_A$ when $t \in (0, T_1]$.

Using now the definition of P in terms of P^0, i.e. (4.4.4) :

$$P(\theta_{T_0} \in A) = \frac{1}{E^0[T_1]} E^0[T_1 \, 1_A].$$

Therefore

$$E^0[T_1 \, 1_A] = E_N^0[T_1 \, 1_A], \quad (A \in F).$$

Since $T_1 > 0$ by construction, $P^0 \equiv P_N^0$.

5 EXAMPLES

5.1 Palm probability and superposition of independent point processes

The situation is that described in Example 1.2.2 with the additional assumption

$$0 < \lambda_i < \infty, \qquad (1 \le i \le k), \tag{5.1.1}$$

where λ_i is the intensity of N_i. Recall that for each $1 \le i \le k$, $(\Phi_i, T_t^{(i)}, \mathbf{P}_i)$ is a stationary point process. We will denote by P_i^0 the associated Palm probability. We are going to prove the following formula :

$$P_N^0 = \sum_{i=1}^{k} \frac{\lambda_i}{\lambda} \left(\underset{k=1}{\overset{i-1}{\times}} \mathbf{P}_k \right) \times \mathbf{P}_i^0 \times \left(\underset{k=i+1}{\overset{k}{\times}} \mathbf{P}_k \right), \qquad (5.1.2)$$

where $\lambda = \sum_{i=1}^{k} \lambda_i$ is the intensity of (N, θ_t, P), the "sum" of the independent point processes $(\Phi_i, T_t^{(i)}, \mathbf{P}_i)$ $(1 \le i \le k)$.

Indeed, by definition, for $A = \prod_{i=1}^{k} A_i$ where $A_i \in \mathbf{M}_i$ $(i=1,k)$,

$$P_N^0(A) = \frac{1}{\lambda} E\left[\int_{(0,1]} 1_A \circ \theta_s \, N(ds) \right]$$

$$= \frac{1}{\lambda} \int_{M_1} \dots \int_{M_k} \left\{ \sum_{j=1}^{k} \int_{(0,1]} \left(\prod_{i=1}^{k} 1_{A_i} \circ T_t^{(i)} \right) \Phi_j(dt) \right\} \mathbf{P}_1(dm_1) \dots \mathbf{P}_k(dm_k).$$

But by Fubini's theorem

$$\frac{1}{\lambda_j} \int_{M_1} \dots \int_{M_k} \left\{ \int_{(0,1]} \prod_{j=1}^{k} \left(1_{A_j} \circ T_t^{(j)}\right) \Phi_i(dt) \right\} \mathbf{P}_1(dm_1) \dots \mathbf{P}_k(dm_k) =$$

$$\mathbf{P}_j^0(A_j) \prod_{\substack{i=1 \\ i \ne j}}^{k} P_i(A_i).$$

Therefore

$$P_N^0\left(\prod_{i=1}^{k} A_i \right) = \sum_{i=1}^{k} \frac{\lambda_i}{\lambda} \mathbf{P}_i^0(A_i) \prod_{\substack{j=1 \\ j \ne i}}^{k} P_j(A_j),$$

which implies (5.1.2).

<u>Remark</u> : The interpretation of (5.1.2) is as follows : there is a probability $\frac{\lambda_i}{\lambda}$ that a point of the superposition comes from the ith point process. Also $\left(\bigotimes_{k=1}^{i-1} \mathbf{P}_k \right) \otimes \mathbf{P}_i^0 \otimes \left(\bigotimes_{k=i+1}^{k} \mathbf{P}_k \right)$ is the Palm probability of P associated with the ith point process (check).

5.2 Palm probability associated with selected marks

Let (N, θ_t, P) be a stationary point process with finite intensity, and for $U \in F$, define

$$N_U(\omega, C) = \int_C 1_U(\theta_t(\omega)) N(\omega, dt), \quad C \in B.$$

For instance, if $Z=(Z_n, n \in \mathbb{Z})$ is a sequence of marks associated with (N,θ_t,P), with values in $(K,\mathcal{K})$ one could take

$$U = \{Z_0 \in L\}$$

for some $L \in \mathcal{K}$. Then, since $Z_0(\theta_{T_n}(\omega)) = Z_n(\omega)$,

$$N_U(C) = \sum_{n\in\mathbb{Z}} 1_L(Z_n)\, 1_C(T_n).$$

In this case N_U counts the points of T_n with a mark Z_n falling in C.

(N_U,θ_t,P) is obviously a stationary point process with finite intensity (since $N_U \leq N$). Its Palm probability is given by

$$P^0_{N_U}(A) = \frac{1}{\lambda_U} E\Big[\int_{(0,1]} 1_A\circ\theta_s\, N_U(ds)\Big], \tag{5.2.2}$$

where λ_U is the intensity of N_U. But

$$\lambda_U = E\Big[\int_{(0,1]} 1_U\circ\theta_t\, N(dt)\Big] = P^0_N(U).$$

Therefore, since $E\Big[\int_{(0,1]} 1_A\circ\theta_s\, N_U(ds)\Big] = E\Big[\int_{(0,1]} 1_A\circ\theta_s\, 1_U\circ\theta_s\, N(ds)\Big] = \lambda P^0_N(A\cap U)$

$$P^0_{N_U}(A) = \frac{P^0_N(A\cap U)}{P^0_N(U)},$$

that is to say

$$P^0_{N_U}(A) = P^0_N(A\,|\,U). \tag{5.2.3}$$

5.3 Palm probabilities associated with selected transitions of a stationary Markov chain

The setting and the notations of the present subsection are the same as those of subsection 1.3, example 3.

The intensity λ_H of N_H is

$$E[N_H(0,1]] = E\Big[\int_{(0,1]} 1_H(X_{s-},X_s)\, N(ds)\Big] =$$

$$= E\Big[\int_{(0,1]} \sum_{(i,j)\in H} 1_{\{i\}}(X_{s-})\, N_{ij}\,(ds)\Big]$$

where $N_{ij} = N_{\{(i,j)\}}$. Therefore from Levy's formula

$$\lambda_H = E[\int_{(0,1]} \sum_{(i,j)\in H} 1_{\{i\}}(X_{s-})\, q_{ij} ds] = \int_{(0,1]} \sum_{(i,j)\in H} P(X_s = i) q_{ij}\, ds$$

that is to say

$$\lambda_H = \sum_{(i,j)\in H} \pi(i)\, q_{ij}. \tag{5.3.3}$$

We assume that

$$0 < \lambda_H < \infty, \tag{5.3.4}$$

so that one can define the Palm probability $P^0_{N_H}$ associated with N_H.

Let now $g : E\times E \to \mathbb{R}$ be non-negative and measurable.

Then

$$E^0_{N_H}[g(X\ ,X\ )] = \frac{1}{\lambda_H} E[\int_{(0,1]} g(X_{s-},X_s)\, N_H(ds)]$$

$$= \frac{1}{\lambda_H} E[\int_{(0,1]} \sum_{(i,j)\in H} g(i,j)\, 1_{\{i\}}(X_{s-})\, N_{ij}\,(ds)]$$

$$= \frac{1}{\lambda_H} E[\int_{(0,1]} \sum_{(i,j)\in H} g(i,j)\, 1_{\{i\}}(X_{s-})\, q_{ij}\, ds],$$

so that finally :

$$E^0_{N_H}(g(X\ _-,X\ )) = \frac{\sum_{(i,j)\in H} g(i,j)\pi(i) q_{ij}}{\sum_{(i,j)\in H} \pi(i)\, q_{ij}}\ . \tag{5.3.5}$$

In particular, we have

$$P^0_{N_H}(X_0 = k) = \frac{\sum_{i/(i,k)\in H} \pi(i)\, q_{ik}}{\sum_{(i,j)\in H} \pi(i)\, q_{ij}}\ . \tag{5.3.6}$$

6 LOCAL ASPECT OF PALM PROBABILITY

6.1 Korolyuk-Dobrushin's infinitesimal estimates

Let (N, θ_t, P) be a stationary point process with finite intensity λ, and let P^0_N be the associated Palm probability. The inversion formula gives

$$P(N((0,t]) > 1) = P(T_2 \le t) = \lambda \int_0^{\infty} P_N^0(u < -T_{-1},\ T_2 \circ \theta_{-u} \le t)du.$$

But $u < -T_{-1}$ implies $T_2 \circ \theta_{-u} = T_1 + u$ on Ω_0. Therefore

$$P(N((0,t]) > 1) = \lambda \int_0^{t} P_N^0(u < -T_{-1},\ T_1 \le t-u)du \le \lambda \int_0^{t} P_N^0(T_1 \le t)du.$$

Since $P_N^0(T_1 > 0) = 0$, this implies

$$P(N((0,t]) > 1) = 0(t), \tag{6.1.1}$$

where $\lim_{t \downarrow 0} 0(t)/t = 0$. This relation is attributed to Korolyuk. One can similarly obtain Dobrushin's relation

$$P(N((0,t]) > 0) = \lambda t + 0(t). \tag{6.1.2}$$

Indeed

$$P(N((0,t]) > 0) = P(T_1 \le t) = \lambda \int_0^{\infty} P_N^0(u < -T_{-1}, T_1 \circ \theta_{-u} \le t)du =$$

$$\lambda \int_0^{t} P_N^0(u < -T_{-1})du = \lambda t + 0(t).$$

6.2 Conditioning at a point

The local interpretation of Palm probability is contained in the following result

$$\lim_{t \downarrow 0} \sup_{A \in F} \left| P_N^0(A) - P(\theta_{T_1} \in A \mid T_1 \le t) \right| = 0. \tag{6.2.1}$$

Proof :

$$P(\theta_{T_1} \in A \mid T_1 \le t) = \frac{\lambda t}{P(T_1 \le t)} \, \frac{P(T_1 \le t, \theta_{T_1} \in A)}{\lambda t}.$$

From Dobrushin's theorem, $\lim_{t \downarrow 0} \frac{\lambda t}{P(T_1 \le t)} = 1$. Also

$$P(T_1 \le t,\ \theta_{T_1} \in A) = \lambda \int_0^{\infty} P_N^0(u < -T_{-1},\ T_1 \circ \theta_{-u} \le t,\ \theta_{T_1} \circ \theta_{-u} \in A)du$$

$$= \lambda \int_0^{t} P_N^0(u < -T_{-1},\ A)du,$$

since $\theta_{T_1} \circ \theta_{-u} = \mathrm{id}$ on $\Omega_0 \quad \{u < -T_{-1}\}$ when $u > 0$. Moreover

$$\lambda t P_N^0(A) = \lambda \int_0^t P_N^0(A)\,du.$$

Therefore

$$\lambda t \mid P_N^0(\theta_{T_1} \in A,\ T_1 \le t) - P_N^0(A) \mid = \lambda \int_0^t P_N^0(u \ge -T_{-1}, A)\,du$$

$$\le \lambda \int_0^t P_N^0(u \ge -T_{-1})\,du = O(t),$$

where the function $t \to O(t)$ does not depend upon A.

We will give in a special case a formulation of (6.2.1) which is more appealing to intuition. Define

$$J(t) = \inf\{h > 0 \mid N(t,t+h)] = 1\}. \tag{6.2.2}$$

We can write

$$P(\theta_{T_1} \in A \mid T_1 \le h) = P(\theta_{J(0)} \in A \mid N(0,h] \ge 1).$$

From the $\theta_{\bullet}$ stationarity of P :

$$P(\theta_{J(0)} \in A \mid N(0,h] \ge 1) = P(\theta_{J(t)} \in A \mid N(t,t+h] \ge 1).$$

For any process $(X(t),\ t \in \mathbb{R})$ such that $X(t,\omega) = X(0, \theta_t\omega)$, and any $C \in B$, this remark will allow us to write

$$P_N^0(X(0) \in C) = \lim_{h \downarrow 0} P(X(J(t)) \in C \mid N(t,t+h] \ge 1). \tag{6.2.3}$$

Also if $(X(t),\ t \in \mathbb{R})$ is right continuous with left hand limits :

$$P_N^0(\Delta X(0) \in C) = \lim_{h \downarrow 0} P(\Delta X(J(t)) \in C \mid N(t,t+h] \ge 1), \tag{6.2.4}$$

where $\Delta X(t) = X(t) - (X(t-)$.

A typical use of (6.2.3) arises in Queueing Theory, when one wants to compute, in a stationary system, the law of the number of customers in the system given that some event (departure, or arrival) occurred.

7 CARACTERIZATION OF POISSON PROCESSES

7.1 Predictable σ-fields

Let $(\mathbf{F}_t, t \in \mathbb{R})$ be a history on $(\Omega, \mathbf{F})$, i.e. an increasing family of sub σ-fields of $\mathbf{F}$. Suppose moreover that

$$\theta_t \mathbf{F}_s = \mathbf{F} \qquad (s,t \in \mathbf{R}). \tag{7.1.1}$$

For instance, if Z is a sequence of marks with values in $(K,\mathbf{K})$, $\mathbf{F}_t = \mathbf{F}_t^{(N,Z)}$ verifies (7.1.1) where $\mathbf{F}_t^{(N,Z)} := \sigma(N_Z(C\times L)\ ;\ L \in \mathbf{K},\ C \in B),\ C \subset (-\infty,t])$.

Define on $\Omega \times \mathbf{R}$ the two following σ-fields : $P(F_t)$, generated by the sets $A_{s_1}\times(s_1,s_2]$, $s_1,s_2 \in \mathbf{R}$, $A_{s_1} \in F_{s_1}$ and $\theta(\mathbf{F}_t)$, generated by the mappings $(\omega,t) \to v(\theta_t(\omega),t)$ where $v : \Omega \times \mathbf{R} \to \mathbf{R}$ is $\mathbf{F} \times \mathbf{B}$ measurable and for each $t \in \mathbf{R}$, $\omega \to v(\omega,t)$ is F_{0-}-measurable (where we use the notation $\mathbf{F}_{t-} = \bigvee_{s<t} F_s$).

We shall show that

$$P(\mathbf{F}_t^{(N,Z)}) \subset \theta(\mathbf{F}_t^{(N,Z)}) \tag{7.1.2}$$

and that

$$\theta(\mathbf{F}_t^{(N,Z)}) \subset P(\mathbf{F}_t^{(N,Z)}), \tag{7.1.3}$$

so that

$$P(\mathbf{F}_t^{(N,Z)}) \equiv \theta(\mathbf{F}_t^{(N,Z)}). \tag{7.1.4}$$

Proof of (7.1.2) : it suffices to observe that if $s_1,s_2 \in \mathbf{R}$, $s_1 \le s_2$, and if H_{s_1} is a F_{s_1}-measurable random variable, then $v(\omega,t) = H_{s_1}(\theta_{-t}(\omega))\, 1_{(s_1,s_2]}$ is such that for all $t,\omega \to v(\omega,t)$ is F_{0-}-measurable. Indeed, if $t \notin (s_1,s_2]$, $v(\omega,t) = 0$ and if $t \in (s_1,s_2]$, $H_{s_1} \circ \theta_{-t}$ is F_{s_1-t}-measurable and therefore F_{0-}-measurable.

Proof of (7.1.3) : it suffices to show that if $v : \Omega \to \mathbf{R}$ is F_{0-}-measurable, $\omega,t \to v(\theta_t(\omega))$ is $P(\mathbf{F}_t)$ measurable. It is sufficient to do this for all v of the form $\phi(N([a,b)\times L))$, $a \le b \le 0$, $L \in \mathbf{K}$. But in this case $(v\circ\theta_t, t \in \mathbf{R})$ is a left-continuous process adapted to $(F_t^{N,Z}, t \in \mathbf{R})$ and therefore it is F_t-predictable*.

7.2 Stochastic intensity and Radon-Nikodym derivatives

Suppose that $((N,Z),\theta_t,P)$ is a stationary marked point process with marks in (K,K), and that the intensity $\lambda = E[N(0,1]]$ is strictly positive and finite. Let P_N^0 be the Palm probability associated with (N,θ_t,P).

*One says that a process $(X_t, t \in \mathbf{R})$ is adapted to $(\mathbf{F}_t, t \in \mathbf{R})$ if for each $t \in \mathbf{R}$, X_t is F_t-measurable, and that it is F_t-predictable if $(\omega,t) \to X_t(\omega)$ is $P(\mathbf{F}_t)$-measurable. From the definition of $P(\mathbf{F}_t)$ it is immediate that if $(X_t, t \in \mathbf{R})$ is left continuous and adapted to $(F_t, t \in \mathbf{R})$, it is F_t-predictable.

Suppose that $P_N^0 << P$ on $F_{0-}^{(N,Z)}$ and define :

$$\tilde{\mu}_0 = \frac{dP_N^0}{dP} \mid F_{0-}^{(N,Z)}. \tag{7.2.1}$$

Then

$$\lambda_t = \lambda . \tilde{\mu} o \theta_t \tag{7.2.2}$$

is a $F_t^{(N,Z)}$ stochastic intensity of (N,P), i.e. :

$$(\lambda_t, t \in \mathbf{R}) \text{ is adapted to } (\mathbf{F}_t^{(N,Z)}, t \in \mathbf{R}) \tag{7.2.3}$$

and

$$E[\int_{\mathbf{R}} X_t \, N(dt)] = E[\int_{\mathbf{R}} X_t \, \lambda_t \, dt], \tag{7.2.4}$$

for all non-negative F_t-predictable processes $(X_t, t \in \mathbf{R})$.

<u>Proof</u> : Let $v : \Omega \times \mathbf{O} \to \mathbf{R}_+$ be $\mathbf{B} \times \mathbf{F}$ measurable and such that for each $t \in \mathbf{R}$, $\omega \mapsto v(\omega,t)$ is $F_{0-}^{(Z,N)}$-measurable. Then

$$\int_\Omega \int_{\mathbf{R}} v(\omega,t) \, P_N^0(d\omega)dt = \int_\Omega \int_{\mathbf{R}} v(\omega,t) \, \tilde{\mu}(\omega) \, P(d\omega)dt$$

$$= \int_\Omega \int_{\mathbf{R}} v(\theta_t \omega,t) \, \tilde{\mu}(\theta_t \omega) \, P(d\omega)dt,$$

where we have used definition (7.2.1) and the θ_t-invariance of P. Therefore, the generalized Campbell's formula reads

$$\int_\Omega \int_{\mathbf{R}} v(\theta_t \omega,t) \, \lambda . \tilde{\mu}(\theta_t \omega)dt \, P(d\omega) = \int_\Omega \int_{\mathbf{R}} v(\theta_t \omega,t) \, N(dt) \, P(d\omega)$$

and (7.2.4) then follows by (7.1.4). □

Conversely, suppose now that $(\lambda_t, t \in \mathbf{R})$ is the $F_t^{(N,Z)}$-intensity of (N,P) where

$$\lambda_t = \lambda . \tilde{\mu} o \theta_t \tag{7.2.5}$$

and $\tilde{\mu}$ is a $F_{0-}^{(N,Z)}$ measurable random variable. Then $P_N^0 << P$ on $F_{0-}^{(N,Z)}$ and (7.2.1) holds.

<u>Proof</u> : Let f be a non-negative $F_{0-}^{(N,Z)}$-measurable random variable. Since $(X_t, t \in \mathbf{R})$ defined by $X_t = f o \theta_t$ is F_t-predictable,

$$\lambda \, E[\int_{(0,1]} f o \theta_s \, \mu o \theta_s \, ds] = E[\int_{(0,1]} f o \theta_s \, N(ds)].$$

But $E[\int_{(0,1]} f\circ\theta_s \, N(ds)] = \lambda \, E_N^0[f]$ and $E[\int_{(0,1]} f\circ\theta_s . \mu\circ\theta_\bullet \, ds] = E[\int_{(0,1]} f.\tilde{\mu} \, ds] = E[f\tilde{\mu}]$.

Therefore $E_N^0[f] = E[f.\tilde{\mu}]$. □

Remark : It can be proved (see [10]) that predictable intensities must be strictly positive at the points T_n, i.e. $\lambda_{T_n} > 0$ P.a.s. Since $P_N^0 << P$, this imples that $\lambda_{T_N} > 0$ P_N^0-a.s. In particular $\lambda_{T_0} = \lambda\tilde{\mu} > 0$, P.a.s .

This shows that P and P_N^0 must be equivalent if $P_N^0 << P$. □

7.3 Palm view of Watanabe's characterization theorem

(7.3.1) If P and P_N^0 agree on $\mathbf{F}_{0-}^{(N,Z)}$, then (N,P) is a Poisson process and moreover $\mathbf{F}_{a}^{(N,Z)}$ is independent of $N((a,b])$ for all $a,b \in \mathbf{R}$, $a \leq b$.

Proof : According to the results of §7.2, (N,P) admits a $F_t^{(N,Z)}$-intensity which is constant. Therefore, by Watanabe's characterisation theorem, (7.3.1) holds.

Remark : If one is willing to admit that the generating function $\psi : C_k \to \mathbf{R}$ ($C_k = \{f: \mathbf{R} \to \mathbf{R}$; f bounded with compact support$\}$) defined by

$$\psi(f) = E[\exp\{\int_{\mathbf{R}} f(s) \, N(ds)\}] \tag{7.3.2}$$

characterizes P, and that for a Poisson process with intensity λ

$$\psi(f) = \exp\{\lambda \int_{\mathbf{R}} (e^{f(s)}-1) ds\}, \tag{7.3.3}$$

a direct proof of (7.3.1) is easy : Indeed define

$$L_{a,t}(f) = \exp\{ \int_{(a,t]} f(s) N(ds)\}.$$

We have, for any $A \in F_a$

$$1_A \, L_{a,t}(f) = 1_A + \int_{(a,t]} 1_A \, L_{a,s-}(f) \, (e^{f(s)}-1) \, N(ds).$$

Taking expectations and using (7.2.4) :

$$E[1_A \, L_{a,t}(f)] = 1_A + \int_{(a,t]} E[1_A \, L_{a,s}(f)](e^{f(s)}-1)\lambda \, ds$$

and therefore

$$E[1_A L_{a,t}(f)] = P(A) \exp\{\lambda \int_a^t (e^{f(s)}-1)ds\}. \tag{7.3.4}$$

Letting $A = \Omega$ and $a \downarrow -\infty$, $t \uparrow +\infty$ (recall that $f \in C_k$) we obtain (7.3.3), and therefore (N,P) is Poisson with intensity λ.

Also from (7.3.4)

$$E[1_A \exp\{\int_a^\infty f(s)N(ds)\}] = P(A)\ E[\exp\{\int_a^\infty f(s)N(ds)\}]. \tag{7.3.5}$$

Since this is true for all $A \in F_a^{(N,Z)}$, we see that the restriction of N to $[a,\infty)$ is P-independent of $F_a^{(N,Z)}$. ∎

8 ERGODICITY OF POINT PROCESSES

8.1 Invariant events

Let (N,θ_t,P) be a stationary point process with non zero and finite intensity and let P_N^0 be the associated Palm probability. Denote θ_{T_1} by θ.

The following result holds

(8.1.1) <u>Let $A \in \mathbf{F}$ be such that $\theta_t A = A$ $(t \in \mathbb{R})$. In other words A is θ_t-invariant. Then $P(A) = 1$ if and only if $P_N^0(A) = 1$.</u>

Indeed, suppose $P_N^0(A) = 1$. The inversion formula yields

$$\begin{aligned} P(A) &= \lambda \int_0^\infty P_N^0(u < T_1, \theta_u A)du \\ &= \lambda \int_0^\infty P_N^0(u < T_1, A)du && (\theta_\cdot\text{-invariance of } A) \\ &= \lambda \int_0^\infty P_N^0(u < T_1)du && (P_N^0(A) = 1) \\ &= \lambda\, E_N^0[T_1] = 1. \end{aligned}$$

Conservely, supposing $P(A)=1$, and using part of the above calculation, we get

$$P(A) = 1 = \lambda \int_0^\infty P_N^0(u < T_1, A)du.$$

But since $1 = \lambda \int_0^\infty P_N^0(u < T_1)du$, we obtain

$$0 = \lambda \int_0^\infty P_N^0(u < T_1, \bar{A})du$$

and therefore $P_N^0(u < T_1, \overline{A}) = 0$, du-almost everywhere, from which we conclude that $P_N^0(\overline{A}) = 0$. □

In a similar fashion, we obtain the following result

(8.1.2) Let $A \in \mathcal{F}$ such that $\theta A = A$. In other words A is θ-invariant. Then $P_N^0(A)=1$ if and only if $P(A)=1$.

$$P_N^0(A) = \frac{1}{\lambda t} E\Big[\int_{(0,t]} (1_A \circ \theta_s)\, N(ds)\Big]$$

$$= \frac{1}{\lambda t} E\Big[\int_{(0,t]} 1_A\, N(ds)\Big] \qquad (\theta^n\text{-invariance of A and } \theta_{T_n} = \theta^n \text{ for } n\geq 1)$$

$$= \frac{1}{\lambda t} E[1_A\, N(0,t])]$$

$$= \frac{1}{\lambda t} E[\, N(0,t])] \qquad (P(A) = 1)$$

$$= 1.$$

Conservely supposing $P_N^0(A) = 1$ and using the above calculation, we get

$$\lambda t P_N^0(A) = \lambda t = E[1_A\, N((0,t])]$$

that is

$$E[1_{\overline{A}}\, N((0,t])] = 0, \text{ for all } t \geq 0$$

which implies $P(\overline{A}) = 0$ (recall the assumption $N(R_+) = \infty$). □

8.2 Ergodicity under the stationary probability and its Palm probability

Let (N, θ_t, P) be a stationary point process with a finite intensity λ, and let P_N^0 be its associated Palm probability. Denote $\theta_{T_1} = \theta$.

The result of this subsection is

(8.2.1) (P, θ_t) is ergodic if and only if (P_N^0, θ) is ergodic.

It is a consequence of the characterization of ergodic probabilities as extremal invariant probabilities. For instance

(8.2.2) (P_N^0, θ) is ergodic if and only if there exist no decompostion

$$P_N^0 = \alpha_1 Q_1 + \alpha_2 Q_2 \qquad (\alpha_1 + \alpha_2 = 1,\ \alpha_1 > 0,\ \alpha_2 > 0)$$

where Q_1 and Q_2 are θ-invariant probabilities.

Similarly :

(8.2.3) <u>(P,θ_t) is ergodic if and only if there exist no decomposition</u>

$$P = \beta_1 P_1 + \beta_2 P_2 \qquad (\beta_1+\beta_2 = 1 \; ; \; \beta_1 > 0, \; \beta_2 > 0)$$

<u>where P_1 and P_2 are θ_t-invariant probabilities $(t \in \mathbb{R})$</u>

(cf Billingsley (1968), "Convergence of Probability Measures", Wiley and Sons, N.Y.).

We can now prove the announced result. Suppose for instance that (P,θ_t) is ergodic and that (P_N^0,θ) is not. Then there must be a decomposition of the type (8.2.2). Let P_1 and P_2 be the stationary probabilities associated with Q_1 and Q_2 (see § 4.4)

The inversion formula applied to (8.2.1) yields

$$P = \alpha_1 \frac{\lambda}{\lambda_1} P_1 + \alpha_2 \frac{\lambda}{\lambda_2} P_2,$$

where $1/\lambda = E_N^0[T_1]$, $1/\lambda_1 = E_{Q_1}[T_1]$, $1/\lambda_2 = E_{Q_2}[T_2]$ (note that λ_1 and λ_2 must be strictly positive). Therefore (θ_t,P) is not ergodic, hence a contradiction.

As for the proof of the converse part, it is this time based on the observation that (8.2.2) implies

$$P_N^0 = \beta_1 \frac{\lambda_1}{\lambda} P_{1,N}^0 + \beta_2 \frac{\lambda_2}{\lambda} P_{2,N}^0,$$

where $\lambda = E[N(0,1]]$, $\lambda_1 = E_{P_1}[N(0,1]]$, $\lambda_2 = E_{P_2}[N(0,1]]$. □

8.3 The cross ergodic theorems

Let (N,θ_t,P) be a stationary point process. It is supposed to be ergodic, i.e. (P,θ_t) is ergodic.

Assume its intensity λ to be non zero and finite and let P_N^0 be its associated Palm probability.

Let f be in $L^1(P)$. Since (P,θ_t) is ergodic, the event

$$A = \{ \lim_{T\to\infty} \frac{1}{T} \int_0^T f\circ\theta_t \, dt = E[f] \}$$

is P-almost certain. Moreover it is θ_t-invariant. Therefore by the results of §8.3, $P_N^0(A) = 1$, i.e.

$$\lim_{T\to\infty} \frac{1}{T}\int_0^T f\circ\theta_t \, dt = E[f], \quad P_N^0\text{-a.s.} \tag{8.3.1}$$

Similarly, since (P_N^0, θ) is ergodic

$$B = \{ \lim_{n\to\infty} \frac{1}{n} \sum_{k=1}^{n} f\circ\theta_{T_k} = E_N^0[f]\}$$

is P_N^0-almost certain and θ-invariant and therefore $P(B) = 1$, i.e.

$$\lim_{n\to\infty} \frac{1}{n} \sum_{k=1}^{n} f\circ\theta_{T_k} = E_N^0\,[f], \quad P\text{-a.s.} \tag{8.3.2}$$

The two above results are essential for the stationary queueing system theory in that they allow an easy interpretation of the "cross formulae" of the Little's type (linking P_N^0- means and P- means of operational quantities such as the P_N^0- mean waiting time and the P- mean number of customers). See §2.2 of Part 2.

REFERENCES FOR PART 1 : PALM PROBABILITIES

The two basic references concerning the foundations of point process theory are

[1] Matthes K., Kertan J., Mecke J., (1974)
Unbegrenzt teilbare Punktprozesse, Akademie Verlag, Berlin (english version : Infinitely divisible point processes, Wiley, New York).

and

[2] Neveu J., (1977)
Processus ponctuels, in Ecole d'Eté de Probabilités de Saint Flour VI-1976, Lecture Notes in Maths 598, Springer Verlag, Heidelberg., pp. 249-447.

The first appearance of the idea of conditioning at a point (see §6) is in

[3] Palm C., (1943)
Intensitatschwankugen in Fernsprechverkehr, Ericsson Technics 44, pp.1-198.

The generalization of Palm's idea is due to

[4] Ryll-Nardzewsky C., (1961)
Remarks on processes of calls, in Proceedings of the 4th Berkeley Symposium II, pp.455-465.

The definition of Palm probability chosen in this monograph, as well as the so-called generalized Campbell formula of §3.3 are due to

[5] Mecke J., (1967)
Stationare zufallige Masse und lokalkompakten Abelschen Gruppen, Z.für W.9, pp.36-58.

The cycle formula of §3.4 was given in its general form by

[6] Neveu J., (1976)
Sur les mesures de Palm de deux processus ponctuels stationnaires, Z.für W.34, pp.199-203.

The characterization theorem for Poisson processes in terms of Palm probabilities is due to

[7] Mecke J., (1968)
Eine karakterische Eigenschaft der doppelt stochastishen Poissonschen Prozesse, Z.für W.11, pp.74-81.

The characterization theorem for Poisson processes in terms of martingales is due to

[8] Watanabe K., (1964)
On discontinuous additive functionals and Levy measures of a Markov process. Japanese J. Math. 34, pp. 53,70.

Some ideas in §7 relating stochastic intensity and Radon-Nikodym derivatives of P and P_N^0 restricted to the strict past at 0 are due to

[9] Papangelou F., (1972)
Integrability of expected increments of point processes, ACM Transactions, 165, pp.483-506

See also in connection with the martingale approach to point processes

[10] Brémaud P., (1981)
Point processes and queues : martingale dynamics, Springer-Verlag, New York.

For further information, the reader should consult the bibliographical references of [1] and [2].

PART II

STATIONARY QUEUEING SYSTEMS

1 - THE G/G/1/∞ QUEUE : CONSTRUCTION OF THE CUSTOMER-STATIONARY STATE

1.1 Loynes' problem

Let $(\Omega,\mathbf{F},P^0)$ be a probability space where P^0 is invariant with respect to some measurable automorphism : $(\Omega,\mathbf{F}) \to (\Omega,\mathbf{F})$. Suppose moreover that

$$(P^0,\theta) \text{ is ergodic} \tag{1.1.1}$$

Let σ be a nonnegative and τ a strictly positive random variable on $(\Omega,\mathbf{F},P^0)$. Both τ on σ will be assumed to be integrable w.r.t. P^0. These variable receive the following interpretation : $\sigma_n = \sigma\circ\theta^n$ is the amount of service required by customer n, $n \in \mathbb{Z}$, and $\tau_n = \tau\circ\theta^n$ is the time between the arrivals of customer n and customer n+1. Customer 0 is supposed to arrive at time 0.

Assume that for some n, the workload just before the arrival of customer n is W_n. Since there is a single server providing service at unit rate, the workload found by customer n+1 upon its arrival, W_{n+1} is given by

$$W_{n+1} = (W_n + \sigma_n - \tau_n)^+, \tag{1.1.2}$$

where $a^+ = \max(a,0)$ for $a \in \mathbb{R}$. (See figure 1, where T_n denotes the arrival date of customer n).

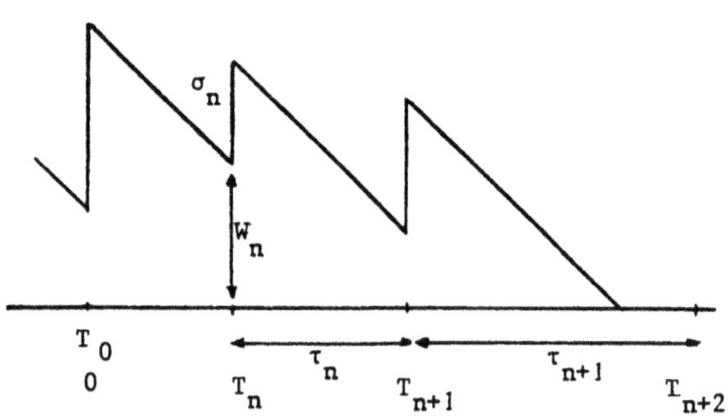

Figure 1

Loynes' problem consists in finding the stationary regime of this workload process. Let us show that this is equivalent to proving existence of a non negative and a.s. finite random variable W on $(\Omega, \mathbf{F}, P^0)$ satisfying the equation

$$W \circ \theta = (W + \sigma - \tau)^+. \tag{1.1.3}$$

If there exists a stationary regime, i.e. a non negative and a.s. finite random variable on $(\Omega, \mathbf{F}, P^0)$ representing the workload found by customer 0, equation (1.1.2) shows that the workload found by customer 1, $W\circ\theta$, should be given by (1.1.3).

Conversely any solution of equation (1.1.3) provides a stationary process $(W_n, n \in \mathbb{Z})$ on $(\Omega, \mathbf{F}, P^0)$ satisfying the recurrence equation (1.1.2) for all $n \in \mathbb{Z}$: Take $W_n = W\circ\theta^n$.

We shall first provide a non negative but possibly infinite solution of (1.1.3). For $n \in \mathbb{N}$, let

$$M_n = \left(\max_{0 \le k \le n} \left\{ \sum_{i=1}^{k} (\sigma\circ\theta^{-i} - \tau\circ\theta^{-i}) \right\} \right)^+, \tag{1.1.4}$$

where $\sum_{i=1}^{0} = 0$ by convention. M_n is the workload found by customer zero when customer $-n$ finds an empty queue upon arrival $(W_{-n} = 0)$. From the very definition, M_n is nondecreasing in n and non negative. One also gets by induction that

$$M_{n+1}\circ\theta = (M_n + \sigma - \tau)^+, \; n \in \mathbb{N}. \tag{1.1.5}$$

Due to increasingness, there exists a non negative but possibly infinite limit

$$M_\infty = \lim \uparrow M_n. \tag{1.1.6}$$

Letting n go to infinity in (1.1.5) one gets

$$M_\infty\circ\theta = (M_\infty + \sigma - \tau)^+. \tag{1.1.7}$$

1.2 Existence of a finite stationary workload

Using the identity $(a-b)^+ = a - a\wedge b$, (1.1.5) takes the form

$$M_{n+1}\circ\theta = M_n - M_n \wedge(\tau-\sigma). \tag{1.2.1}$$

Using now the increasingness of M_n, the θ- invariance of P^0 and the integrability of σ and τ, one gets that $E^0[M_n\wedge(\tau-\sigma)] \leq 0$. By dominated convergence

$$E^0[M_\infty \wedge(\tau-\sigma)] \leq 0. \tag{1.2.2}$$

From (1.1.7), the event M_∞ is θ invariant. Since (P^0,θ) is ergodic, either $M_\infty = \infty$ P^0 - a.s. or $M_\infty < \infty$ P^0 - a.s. If $M_\infty = \infty$ P_0 - a.s., equation (1.2.2) entails that $E^0[\tau-\sigma] \leq 0$. Therefore :

<u>If</u> $E^0[\sigma] < E^0[\tau]$, $W = M_\infty$ <u>is an a.s. finite solution of (1.1.3).</u> (1.2.3)

Let us show conversely that

<u>If</u> $E^0[\sigma] > E^0[\tau]$, <u>there is no finite solution of (1.1.3).</u> (1.2.4)

One first gets that M_∞ is the smallest non negative solution of (1.1.3) : $W \geq M_0 = 0$ and if $W \geq M_n$, $W\circ\theta = (W + \sigma - \tau)^+ \geq (M_n + \sigma-\tau)^+ = M_{n+1}\circ\theta$. Therefore, by induction, $M_n \leq W$ so that $M_\infty \leq W$.

In addition, the ergodic assumption entails

$$\lim_{n\to\infty} \frac{1}{n} \sum_{i=1}^{n} (\sigma\circ\theta^{-i} - \tau\circ\theta^{-i}) = E^0[\sigma-\tau] > 0,$$

so that in view of (1.1.4), $M_\infty = \infty$. ◻

1.3 Uniqueness of the stationary workload

We shall need the following lemma :

<u>If</u> M <u>is non negative and finite and if</u> $M-M\circ\theta \in L^1(P^0)$, <u>then,</u> $E^0[M-M\circ\theta] = 0$.

<u>Proof</u> : For any $c > 0$, $|M\wedge c-(M\wedge c)\circ\theta| \leq |M-M\circ\theta|$. The conclusion follows from $E^0[M\wedge c - (M\wedge c)\circ\theta] = 0$ and dominated convergence. ◻

The problem of uniqueness is partially answered by :

(1.3.1) <u>If</u> $E^0[\sigma] < E^0[\tau]$, M_∞ <u>is the unique non negative and finite solution of (1.1.3).</u>

Indeed, let W be another solution. We know that $W \geq M_\infty$. Hence, $\{W = 0\} \subset \{W = M_\infty\}$. This last event being θ - invariant, if we show that $P^0[W=0] > 0$, the ergodic assumption will then entail that $W = M_\infty$ P^0 - a.s. Suppose that $W > 0$ P^0 - a.s., we get from (1.1.3) that

$$W o\theta = W + \sigma - \tau \quad P^0.\ \text{a.s.} \tag{1.3.2}$$

Accordingly, the lemma entails $E^0[\sigma] = E^0[\tau]$, which contradicts our assumption. □

Let us now assume that there exists a non negative and finite solution of (1.1.3) different from M_∞ (so that necessarily, $E^0[\sigma] = E^0[\tau]$). Since M_∞ is the smallest solution of (1.1.3), $M_\infty < \infty$ P^0 - a.s. Therefore, in view of (1.1.2) and the lemma,

$$0 = E^0[M_\infty \wedge (\tau - \sigma)]$$

$$= E^0[1_{\{M_\infty \geq \tau - \sigma\}}(\tau - \sigma)] + E^0[1_{\{M_\infty < \tau - \sigma\}} M_\infty].$$

Since $E^0[\tau] = E^0[\sigma]$, we have

$$E^0[1_{\{M_\infty < \tau - \sigma\}}(\tau - \sigma)] = E^0[1_{\{M_\infty \geq \tau - \sigma\}}(\tau - \sigma)],$$

so that

$$E^0[1_{\{M_\infty < \tau - \sigma\}}(M_\infty - \tau + \sigma)] = 0.$$

Therefore

$$M_\infty \geq \tau - \sigma \quad P^0 \text{ - a.s. }.$$

This implies $M_\infty o\theta = M_\infty + \sigma - \tau$. Hence, for all constant $c \geq 0$, $(M_\infty + c) o\theta = (M_\infty + c + \sigma - \tau)^+$. Therefore :

(1.3.2) <u>If</u> $E^0[\sigma] = E^0[\tau]$, $M_\infty + c$ <u>is a solution of (1.1.3) fo all</u> $c \geq 0$.

1.4 Construction points

(1.4.1) *If* $E^0[\sigma] < E^0[\tau]$, *then* $P^0[M_\infty = 0] > 0$.

We proved in the preceding section that if $M_\infty < \infty$, then $E^0[M_\infty \wedge (\tau-\sigma)] = 0$. Since M_∞ satisfies (1.1.3), $P^0[M_\infty o\theta = 0] \geq P^0[M_\infty < (\tau-\sigma)]$. If $P^0[M_\infty < \tau-\sigma] = 0$ then $0 = E^0[M_\infty \wedge(\tau-\sigma)] = E^0[\tau-\sigma]$. This contradicts our assumption so that necessarily $P^0[M_\infty < \tau-\sigma] > 0$. Therefore $P^0(M_\infty = 0) = P(M_\infty o\theta = 0) > 0$. □

$n \in \mathbb{Z}$ is called a "construction point" for ω if $M_\infty o\theta^n(\omega) = 0$.

(1.4.2) *If* $E^0[\sigma] < E^0[\tau]$, *there are for* P^0*-almost all* ω *an infinite number of positive (resp. negative) construction points.*

This follows from the ergodic theorem and (1.4.1) :

$$\lim_{n\to\infty} 1_{\{0\}}(M_\infty o\theta^n(\omega)) = \infty \quad P^0 - \text{a.s.} \quad \square$$

In the First In First Out case, one can derive from these construction points the stochastic process $\{X(t),\ t \in \mathbb{R}\}$ of the number of customers in the system, hence the terminology. To this purpose, we first construct the sequences of arrival and departure epochs, $(T_n,\ n \in \mathbb{Z})$ and $(D_n,\ n \in \mathbb{Z})$ as suggested by figure 2 :

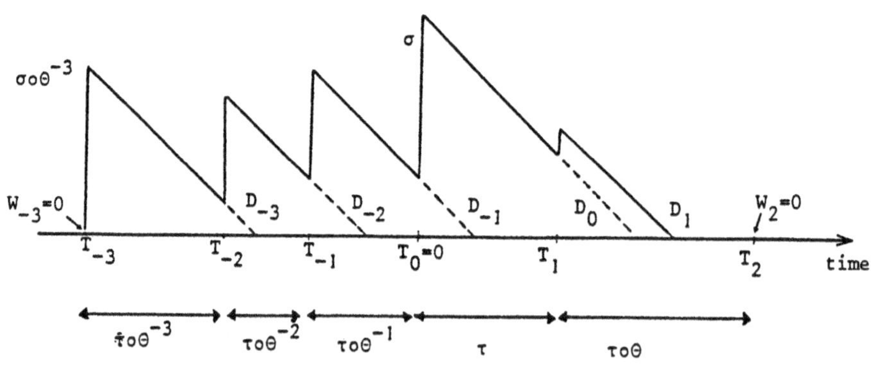

Figure 2

Here, +2 and −3 are construction points and D_n is given by :

$$D_n = T_n + W o \theta^n + \sigma o \theta^n.$$

If n is a construction point, the system is empty just before T_n. One can hence construct X(t) from (T_n, D_n, n ∈ Z/) as indicated in figure 3 :

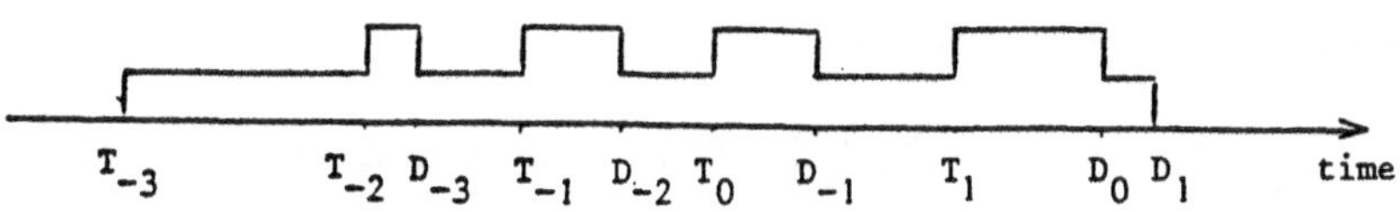

Figure 3

Clearly, (X(t), t ≥ 0) is uniquely defined when adding say a left-continuity assumption. In particular, $X(T_0)$ is the number of customer found by customer 0 upon arrival. We have :

$$\begin{cases} X(T_n) o \theta = X(T_{n+1}),\ n \in Z/, \\ X(D_n) o \theta = X(D_{n+1}),\ n \in Z/. \end{cases} \tag{1.4.3}$$

1.5 Initial workload and long term behaviour

Let Y ≥ 0 be a finite random variable on $(\Omega, \mathbf{F}, P^0)$ representing some initial workload and let (W_n^Y, n ∈ IN) be defined by

$$\begin{cases} W_{n+1}^Y = (W_n^Y + \sigma_n - \tau_n)^+,\ n \in \mathbb{N} \\ W_0^Y = Y. \end{cases} \tag{1.5.1}$$

Let T be the first n such that $W_n^Y = 0$. For $1 \le n < T$, $W_n^Y = Y + \sum_{i=0}^{n-1} (\sigma_i - \tau_i)$. Under the hypothesis $E^0[\sigma] < E^0[\tau]$, the ergodic assumption entails that $\lim \sum_{i=0}^{n} (\sigma_i - \tau_i) = -\infty$. Thus, $T < \infty$ P^0 - a.s.

Consider now the sequence $(W_n^0, n \in \mathbb{N})$ obtained for $Y = 0$. One gets by induction that $W_n^Y \geq W_n^0$, $n \in \mathbb{N}$, so that (1.1.5) entails that $W_n^Y = W_n^0$ for all $n \geq T$ ($W_n^0 \geq W_n^Y = 0$).

(1.5.2) <u>If $E^0[\sigma] < E^0[\tau]$, then, for any finite random variable</u> Y, <u>there exists a finite integer valued random variable</u> T <u>such that for</u> $n \geq T$, $W_n^Y = W_n^0$.

Notice that the sequence W_n^0 is simply related to M_n of (1.1.4) by

$$W_n^0 = M_n \circ \theta^n. \tag{1.5.3}$$

This together with (1.5.2) entails

(1.5.4) <u>If $E^0[\sigma] < E^0[\tau]$, W_n^Y converges in law to the law of M_∞ when</u> n <u>goes to infinity</u>.

In other words, when the queue is stable, W_n^Y will coincide with W_n^0 in a finite time and will converge in law to M_∞.

2 - FORMULAE FOR THE G/G/1/∞ QUEUE

2.1 Construction of the time stationary workload

Let $(\Omega, \mathbf{F}, P)$ be a probability space endowed with a measurable semi-group of automorphisms $(\theta_t, t \in \mathbb{R})$ and a point process N verifying $N \circ \theta_t = T_t N$ $(t \in \mathbb{R})$. Suppose moreover that (P, θ_t) is ergodic. Define $\theta_{T_1 - T_0} = \theta$ $(\theta = \theta_{T_1})$ on Ω^0). Suppose that a non-negative r.v. σ is also defined on (Ω, F).

Assume that the intensity of N is finite, so that there exists a Palm <u>probability</u> P^0 associated with (N, P, θ_t). Suppose moreover that τ is P^0-a.s. positive that $\sigma, \tau \in L^1(P^0)$ and finally that $E^0\sigma < E^0\tau$.

For <u>all</u> $\omega \in \Omega$ construct $W(\omega)$ by

$$\begin{cases} W(\omega) = M_\infty(\omega) = \lim_{n \to \infty} M_n(\omega) \\ M_n(\omega) = \left(\max_{0 \leq k \leq n} \left\{ \sum_{i=1}^{k} ((\sigma \circ \theta^{-i})(\omega) - (\tau \circ \theta^{-i})(\omega)) \right\} \right)^+ \end{cases} \tag{2.1.1}$$

As we saw in §1, W is the P^0-a.s. unique, P^0-a.s. finite r.v. which verifies

$$W \circ \theta = (W + \sigma - \tau)^+, \quad P^0\text{-a.s.} \tag{2.1.2}$$

Define, for <u>all</u> $\omega \in \Omega$,

$$W_n(\omega) = (W \circ \theta^n)(\omega) \tag{2.1.3}$$

The event

$$A = \{W_n < \infty, W_{n+1} = (W_n + \sigma_n - \tau_n)^+\} \tag{2.1.4}$$

is, by the construction (2.1.3), θ-invariant, and moreover $P^0(A) = 1$. Therefore by the result (8.1.1) of Part 1, $P(A) = 1$, i.e.

$$(2.1.5) \qquad \begin{aligned} &W_n < \infty \quad P\text{-a.s.} \\ &W_{n+1} = (W_n + \sigma_n - \tau_n)^+, \quad P\text{-a.s.} \end{aligned}$$

In particular, if we define for all $t \in [T_n, T_{n+1})$

$$W(t) = (W_n + \sigma_n + T_n - Z^{+} \qquad (2.1.6)$$

we find from (2.1.5), (2.1.6) and the finiteness of σ_n that

$$(2.1.7) \qquad \begin{aligned} &W(t) < \infty, \quad P\text{-a.s.} \\ &W(T_n^-) = W_n, \quad P\text{-a.s.} \end{aligned}$$

Also from the very construction of $W(t)$

$$W(t+s) = W(t) \circ \theta_s \text{ for all } \omega \in \Omega, s,t \in \mathbb{R} \qquad (2.1.8)$$

so that $(W(t), t \in \mathbb{R})$ is a θ_t-stationary left-continuous process on (Ω, F, P) (see Fig. 4).

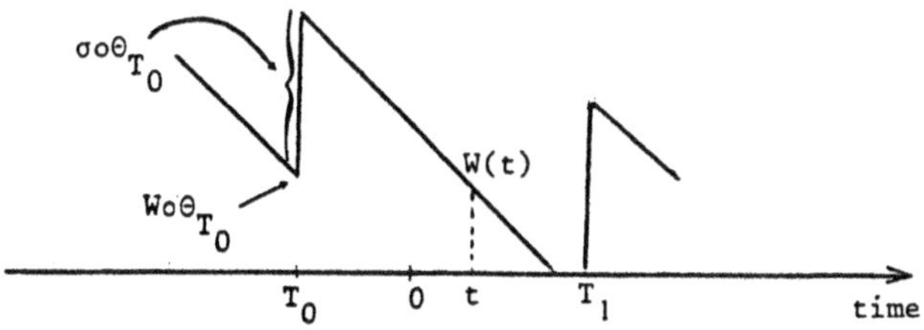

Figure 4

2.2 Little's formula : the FIFO case

Following the lines of section (1.4), in the FIFO case, we can construct a left-continuous stochastic process $(X(t), t \in \mathbb{R})$ on $(\Omega, \mathbf{F}, P)$ representing the number of customers in the queue at time t and such that $X(t+s) = X(t) \circ \theta_s$ for all $t,s \in \mathbb{R}$.

For $x,y \in \mathbb{R}$, let $f(x,y) = 1_{(x \leq 0,\ y \geq -x)}$ and let $\tilde{W}_n = (W+\sigma) \circ \theta_n$. Owing to the FIFO assumption,

$$X(0) = \sum_{n \in \mathbb{Z}} f(T_n, \tilde{W}_n). \qquad (2.2.1)$$

Therefore, the original Campbell's formula (3.3.2), Part 1 and formula (3.1.4), Part 1 entail

$$\begin{aligned} E[X(0)] &= \lambda \int_{\mathbb{R}} \int_{\mathbb{R}_+} f(t,y)\,dt\, P^0[\tilde{W}_0 \in dy] \\ &= \lambda \int_0^\infty (1 - P^0(\tilde{W}_0 \leq t)\,dt, \end{aligned}$$

i.e.

$$E[X(0)] = \lambda E^0[\tilde{W}_0] = \lambda(E^0[W] + E^0[\sigma]). \qquad (2.2.2)$$

This is Little's formula. Since (P,θ) and (P^0,θ) are ergodic, the cross ergodic theorems of section 8.3 (part 1) yield

$$E[X(0)] = \lambda \lim_{n\to\infty} \frac{1}{n} \sum_{k=1}^{n} \tilde{W}(T_n) \qquad P-\text{a.s.} \tag{2.2.2'}$$

and

$$\lambda E^0[\tilde{W}_0] = \lim_{t\to\infty} \frac{1}{t} \int_0^t X(s)ds \qquad P^0 \text{ a.s.} \tag{2.2.2''}$$

2.3 Probability of emptiness

The following "evolution formula" holds for the workload in G/G/1/∞ queues :

$$W(t) = W(0) + \sum_{n\in Z/} \sigma_n 1_{\{0<T_n\le t\}} - \int_0^t (1 - 1_{\{W(s)=0\}})ds. \tag{2.3.1}$$

Indeed, the workload decreases at rate 1 as long as the queue is not empty. Assume stability i.e.

$$\rho = \frac{E^0[\sigma]}{E^0[\tau]} < 1. \tag{2.3.2}$$

W(t) is then P - a.s. finite so that owing to the lemma of § 1.3,

$$E[\sum_{n\in Z/} 1_{\{0<T_n\le t\}} \sigma_n] = E[\int_0^t (1 - 1_{\{X(s)=0\}})ds.$$

Furthermore

$$E[\sum_{n\in Z/} 1_{\{0<T_n\le t\}}\sigma_n] = \lambda t\ E^0[\sigma] = \rho t,$$

so that

$$P[W(t) = 0] = P[X(t) = 0] = 1 - \rho. \tag{2.3.3}$$

Remark. Formula (2.3.3) is true whatever the service discipline (FIFO, LIFO, RANDOM, etc.). The only requirement is that of a work conserving discipline i.e. the server cannot be idle if the system is not empty and the service rate is always 1.

2.4 Takacs' formulae

The evolution equation of the preceding section yields the following formula, valid for all complex numbers s with a positive real part :

$$e^{-sW(t)} = e^{-sW(0)} + \sum_{n \in \mathbb{Z}/} e^{-sW(T_n^-)} (e^{-s\sigma_n} - 1)\, 1_{\{0<T_n \leq t\}} + \\ + s \int_0^t e^{-sW(u)} (1 - 1_{\{X(u) = 0\}}) du. \tag{2.4.1}$$

Therefore, taking expectation with respect to P, observing that $e^{-sW(s)} 1_{\{X(s)=0\}} = 1_{\{X(s)=0\}}$ and using the preceding result (2.3.3), we get :

$$\lambda \cdot E^0[e^{-s(W+\sigma)}] - E^0[e^{-sW}] = s(\rho - E[e^{-sW(0)}]). \tag{2.4.2}$$

This general relation between the Laplace transforms of the time and customer-stationary workload was obtained by L. Takacs. In the particular case of $G/GI/1/\infty$ queue, i.e. a queue where $(\sigma_n, n \in \mathbb{Z}/)$ is an i.i.d. sequence independent of the arrival process, formula (2.4.2) becomes

$$\lambda E^0[e^{-sW}](E^0[e^{-s\sigma}]-1) = s(\rho - E[e^{-sW(0)}]. \tag{2.4.3}$$

3 - THE $G/G/s/\infty$ QUEUE

3.1 The ordered workload vector

The input process is defined on the Palm space of a stationary point process as for $G/G/1/\infty$ queues. There are $s \geq 1$ servers attending customers and the allocation rule is that an arriving customer is assigned the server with the smallest workload. Once assigned, this customer will wait and then be served at unit rate until completion.

We will construct the stationary workload process $(\vec{W}_n, n \in \mathbb{Z}/)$ where $\vec{W}_n = (W_n^1, \ldots, W_n^s)$ is a permutation in increasing order of the workload found in each server by the n-th customer upon arrival : $W_n^1 \leq W_n^2 \leq \ldots \leq W_n^s$ for all $n \in \mathbb{Z}/$.

This ordered vector satisfies the recurrence relation

$$\vec{W}_{n+1} = R(\vec{W}_n + \sigma_n \vec{e} - \tau_n \vec{i})^+, \tag{3.1.1}$$

where $\vec{e} = (1, 0, \ldots, 0)$, $\vec{i} = (1, \ldots, 1)$, R is the operator arranging vectors of $\mathbb{R}^s$ in increasing order, and $(x_1, \ldots, x_s)^+ = (x_1^+, \ldots, x_s^+)$.

The equation to be solved on this Palm space therefore reads

$$\vec{W}o\theta = R(\vec{W} + \sigma\vec{e} - \tau\vec{1})^{+}. \tag{3.1.2}$$

We shall denote by $(\vec{W}_n^Y, n \geq 0)$ the solution of (3.1.1) with initial workload $\vec{Y} \in \mathbb{R}^{+s}$. Define

$$\vec{M}_n = \vec{W}_n^0 o\theta^{-n}. \tag{3.1.3}$$

$\vec{M}_n$ is the ordered workload vector found by customer 0 if customer $-n$ finds an empty system upon arrival. We have

$$\vec{M}_{n+1} o\theta = R(\vec{M}_n + \sigma\vec{e} - \tau\vec{1})^{+}, \ \forall n \geq 0. \tag{3.1.4}$$

From this, we prove by induction that $\vec{M}_n$ increases coordinatewise to a limit $\vec{M}_\infty$ when $n \to \infty$ and that $\vec{M}_\infty$ satisfies (3.1.2). Notice that M_∞ has possibly infinite coordinates.

3.2 Existence of a finite stationary workload vector

Let M_n^j, $j = 1, 0$ be the j-th coordinate of $\vec{M}_n$:

$$M_n^1 \leq M_n^2 \leq \ldots \leq M_n^s, \ \forall n \in \mathbb{N}. \tag{3.2.1}$$

Using the identity $(a-b)^+ = a - a \wedge b$ in (3.1.4), we get

$$\begin{aligned} \sum_{j=0}^{s} M_{n+1}^j o\theta &= (M_n^1 + \sigma - \tau)^+ + \sum_{j=2}^{s} (M_n^j - \tau)^+ \\ &= \sum_{j=1}^{s} M_n^j - (M_n^1 \wedge(\tau-\sigma) + \sum_{j=2}^{s} M_n^j \wedge\tau). \end{aligned} \tag{3.2.2}$$

The lemma of section 1.3 entails

$$E^0[M_n^1 \wedge(\tau-\sigma) + \sum_{j=2}^{s} M_n^j \wedge\tau] \leq 0, \tag{3.2.3}$$

i.e., letting $n \to \infty$,

$$E^0[M_\infty^1 \wedge(\tau-\sigma) + \sum_{j=2}^{s} M_\infty^j \wedge\tau] \leq 0. \tag{3.2.4}$$

Since $\vec{M}_\infty$ satisfies (3.1.2), we have :

$$M_\infty^1 o\theta = (M_\infty^1 + \sigma - \tau)^+ \wedge (M_\infty^2 - \tau)^+ \qquad (3.2.5)$$

and

$$M_\infty^s o\theta = (M_\infty^1 + \sigma - \tau)^+ \vee (M_\infty^s - \tau)^+. \qquad (3.2.6)$$

Therefore, both $\{M_\infty^1 = \infty\}$ and $\{M_\infty^s = \infty\}$ are θ-invariant events. In particular, either $M_\infty^1 = +\infty$ P^0 - a.s. or $M_\infty^1 < \infty$ P^0 - a.s. In the first case, (3.2.3) entails that $E^0[\sigma] \geq s.E^0[\tau]$. Hence we have proved that if $E^0[\sigma] < s.E^0[\tau]$, then $M_\infty^1 < \infty$ P^0 - a.s. Actually, the following stronger result holds :

(3.2.7) <u>If</u> $E^0[\sigma] < sE^0[\tau]$, <u>then</u> $\vec{M}_\infty < \infty$ P^0 - a.s.

To show this, define $Z_n = (M_n^1 + \sigma - \tau)^+$. Equation (3.1.4) entails

$$M_{n+1}^s o\theta = Z_n \quad (M_n^s - \tau)^+ = M_n^s - (M_n^s - Z_n) \wedge \tau.$$

Since $Z_n \uparrow Z_\infty = (M_\infty^1 + \sigma - \tau)^+$, we get

$$M_{n+1}^s o\theta \leq M_n^s - (M_n^s - Z_\infty) \wedge \tau.$$

Therefore,

$$E^0[(M_\infty^s - Z_\infty)] \wedge \tau] \leq 0,$$

so that the assumption $M_\infty^s \uparrow \infty$, $M_\infty^1 < \infty$ entails that $E^0[\tau] \leq 0$, a contradiction. □

Notice that we have not used the assumption $E^0[\sigma] < sE^0[\tau]$ to get this result. Therefore, the only way for $\vec{M}_\infty$ to be infinite with non zero probability is that $M_\infty^j = \infty$, P_0 - a.s., for all j ($1 \leq j \leq s$). In this case, we find from (3.2.4) that $E^0[\sigma] \geq sE^0[\tau]$. Finally

(3.2.8) <u>If</u> $\vec{M}_\infty$ <u>is not</u> P^0 - a.s. <u>finite, then</u> $E^0[\sigma] \geq sE^0[\tau]$.

3.3 Construction points

Following the lines of section 1.4, one sees that it will be possible to construct the queue length process from the knowledge of $\vec{M}_\infty$ provided there is an infinite number of construction points where $M_\infty^1 o\theta^n = 0$ [Indeed, if $M'_\infty o\theta_1^n = 0$, the number of customers in the system at time $T_n + 0$ is known exactly; it is the number of components of $M_\infty o\theta^n$ which are different from 0]. Let us prove that

(3.3.1) <u>If $E^0[\sigma] < sE^0[\tau]$, there are for P^0 - a.s. all ω an infinity of positive and negative construction points.</u>

Assume that $M^1_\infty \geq \tau - \sigma$ and $M^j_\infty \geq \tau$, $j = 2, \ldots, s$ P^0 - a.s. Then, (3.2.4) entails $E^0[\sigma] \geq sE^0[\tau]$. Hence

$$P^0[M^1_\infty < \tau-\sigma \text{ or } M^j_\infty < \tau \text{ for some } j \geq 2] > 0.$$

In view of (3.1.4) this entails that $P^0[M^j_\infty o\theta = 0 \text{ for some } j] > 0$, which concludes the proof, in view of the ergodicity of (P^0,θ). ▫

The stronger property $P^0[M^s_\infty = 0] > 0$ does not hold for all stable G/G/s/∞ queues as can be checked directly in the following counter example: s = 3 and both σ and τ are deterministic random variables : $\sigma = 2$ and $\tau = 1$.

3.4 The busy cycle formulae

Let N' be the stationary point process counting arrival dates at which the queue is empty

$$N'(C) = \int_C 1_{\{M^s_\infty o\theta_t = 0\}} dN(t) \tag{3.4.1}$$

and assume that its intensity, λ', is strictly positive. Let $(T'_n, n \in \mathbb{Z})$ be the sequence of points of N'. The length of the n-th cycle is

$$C_n = T'_n - T'_{n-1}, \; n \in \mathbb{Z}. \tag{3.4.2}$$

C_n is the sum of an idle period I_n and a busy period B_n

$$C_n = I_n + B_n, \; n \in \mathbb{Z}. \tag{3.4.3}$$

Let also Z_n, $n \in \mathbb{Z}$, $n \in \mathbb{Z}$ be the number of customers served during the n-th cycle :

$$Z_n = N((T'_{n-1}, T'_n]) \tag{3.4.4}$$

and X_n be the number of customers found by customer n upon arrival (see figure 5).

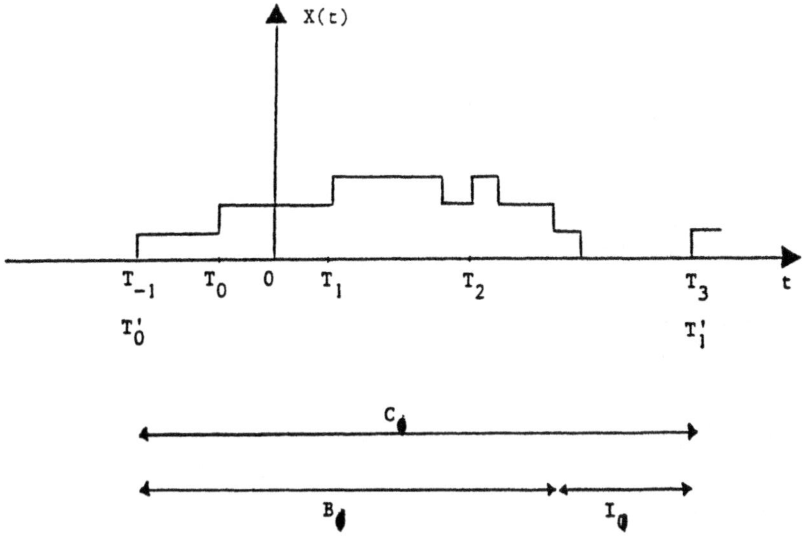

Figure 5

Applying the exchange formula of part 1, § 3.4 to $f = 1_{\{X_0=0\}}$, since $\sum_{n=1}^{N((0,T_1'])} 1_{\{X_0=0\}} = 1$, we get $P^0(X_{0^-} = 0) = \frac{\lambda'}{\lambda}$.

Applying the exchange formula to $f = -T_{-1}$, we obtain

$$\frac{1}{\lambda} = E^0[-T_{-1}] = \frac{E^{0'}\left[\sum_{i=1}^{N((0,T_1'])} -T_{-1}\circ\theta_{T_i}\right]}{E^{0'}[N((0,\ T_1'))]}$$

$$= \frac{E^{0'}(C_1)}{E^{0'}[N((0,\ T_1']))]} ,$$

where $E^{0'}$ denotes integration with respect to the Palm probability of N'.

Therefore, combining the above results,

$$P^0[X_{0^-} = 0] = \frac{E^0[T_1]}{E^{0'}[T_1']} . \tag{3.4.5}$$

where $E^{0'}$ denotes integration with respect to the Palm probability of N'.

From the mean value formula (4.3.2) of part 1,

$$P[X(t) = 0] = \frac{E^{0'}[\int_0^{T'_1} 1_{\{X(t) = 0\}}dt]}{E^{0'}[T'_1]},$$

hence the second cycle formula :

$$P[X(t) = 0] = \frac{E^{0'}[I_1]}{E^{0'}[C_1]}. \tag{3.4.6}$$

4 - THE G/G/1/0 QUEUE

4.1 Definition and examples

The input process $(\sigma_n, \tau_n, n \in \mathbb{Z})$ and the underlying probability space $(\Omega, \mathbf{F}, P^0, \theta)$ are defined in the same terms as in §1.1. There is only one server working at unit rate. But here, the capacity of the working room is 0. Therefore, if a customer finds the server busy upon arrival, he is lost i.e., he leaves immediately the system and never comes back.

The workload W_n found by the n-th potential customer upon arrival, satisfies the recurrence equation :

$$W_{n+1} = (W_n + \sigma_n . 1_{\{W_n = 0\}} - \tau_n)^+, \; n \in \mathbb{Z}. \tag{4.1.1}$$

The corresponding stationary equation is therefore

$$W o \theta = (W + \sigma . 1_{\{W=0\}} - \tau)^+. \tag{4.1.2}$$

The basic question concerning this equation is that of <u>existence</u> of a solution.

Let us see by means of examples that 3 situations can arise : there is a unique solution ; there is a solution but it is not unique ; there is no solution.

(4.1.3) <u>Equation (4.1.2) may have no solution.</u>

Take $\Omega = \{\omega_1, \omega_2\}$, $\theta(\omega_1) = \omega_2$, $\theta(\omega_2) = \omega_1$, $P^0(\{\omega_1\}) = P^0(\{\omega_2\}) = \frac{1}{2}$, $\sigma_1 = 1$, $\sigma_2 > 2$. Suppose that there exists a solution of (4.1.2) then, with obvious notations :

(a) $\quad W_1 = (W_2 + \sigma_2 . 1_{\{W_2=0\}} - 1)^+$

(b) $\quad W_2 = (W_1 + \sigma_1 . 1_{\{W_1=0\}} - 1)^+.$

This system has no solution. Indeed if $W_1 = 0$, then from (b), $W_2 = (\sigma_1-1)^+ = 0$. Therefore, from (a), $0 = (\sigma_2-1)^+ = \sigma_2-1 > 0$, a contradiction. Similarly, if $W_2 = 0$ then from (a), $W_1 = (\sigma_2-1)^+ = \sigma_2-1 > 0$. Therefore, from (b), $0 = (\sigma_2-1-1) = \sigma_2-2 > 0$, another contradiction. The only possibility left is that $W_1 > 0$ and $W_2 > 0$. But this means that all customers are rejected. From (a) and (b), we get $W_1 = ((W_1-1)^+ - 1)^+$. But this is impossible if $W_1 > 0$. □

(4.1.4) Equation (4.1.2) may have two finite solutions.

The setting is the same as for (4.1.3) except that σ_1 and σ_2 now verify $1 < \sigma_i \le 2$, $i = 1,2$. One can check that $(W_1 = 0,\ W_2 = \sigma_1-1)$ and $(W_1 = \sigma_2-1,\ W_2 = 0)$ are both solution of (4.1.2). □

If $1 < \sigma_1 \le 2$ and $\sigma_2 > 2$, there is a unique solution $(W_1 = 0,\ W_2 = \sigma-1)$ which corresponds to serving one customer every second arrival : the customers asking for σ_1.

4.2 Construction of an enriched probability space

Concerning uniqueness, nothing can be done in the general case. However, existence is granted provided one accepts to work on a "enriched" measure space containing more than the input process. A possible construction of such an enriched space goes as follows : Define

$$\nu = \inf\{n \ge 1 \ /\ \sigma \le \sum_{k=0}^{n-1} \tau \circ \theta^k\}, \qquad (4.2.1)$$

where $\nu = \infty$ if the set $\{\ldots\}$ is empty. In words, the ν-th customer is the first to be served after customer 0 if this one receives service (see figure 6). Since P^0 is θ-ergodic and not identically 0, $\sum_{k=0}^{\infty} \tau\circ\theta^k = \infty$, P^0 - a.s., so that $\nu < \infty$, P^0 - a.s.

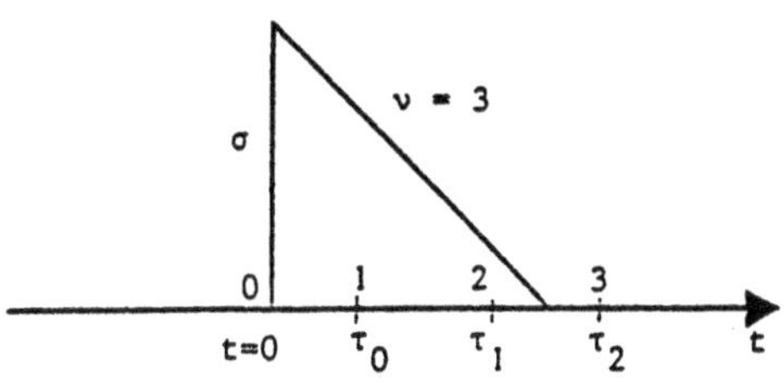

Figure 6

Define

$$B_n = \{\omega \ / \ \forall\ m \geq n,\ \nu\circ\theta^{-m}(\omega) \leq m\}. \qquad (4.2.2)$$

In words, B_n is the set of ω's such that for all $m \geq n$, customer $-m$, whenever he is not rejected, would complete his service before the arrival of customer 0 (see figure 7 where an ω of B_4 is examplified)

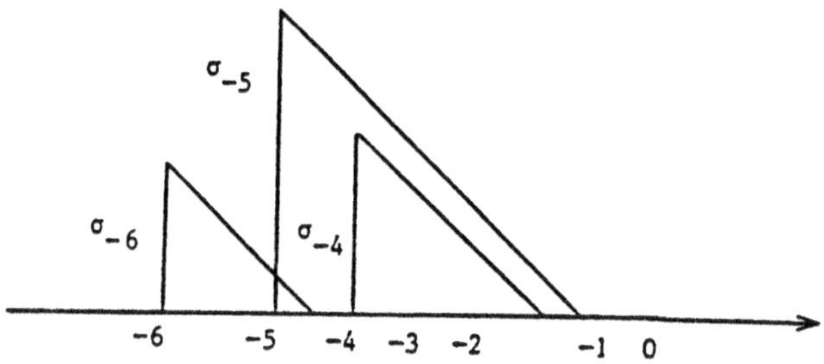

Figure 7

Let us show that

$$\bigcup_{n=1}^{\infty} B_n = \Omega \quad P^0 - \text{a.s.} \qquad (4.2.3)$$

Indeed, $\sum_{k=0}^{\nu-2} \tau\circ\theta^k < \sigma \leq \sum_{k=0}^{\nu-1} \tau\circ\theta^k$ and therefore

$$E^0[\sum_{k=0}^{\infty} \tau\circ\theta^k \, 1_{\{\nu>k+1\}}] < E^0[\sigma].$$

Owing to the θ-invariance of P^0, this reads

$$E^0[\tau\circ\theta^{-1} \sum_{k=1}^{\infty} 1_{\{\nu\circ\theta^{-k}>k\}}] < E^0[\sigma].$$

Since $E^0[\sigma] < \infty$, this entails that $\sum_{k=1}^{\infty} 1_{\{\nu\circ\theta^{-k}>k\}}$ is P^0 - a.s. finite. □

For $\omega \in \Omega$, let $L(\omega) : \mathbb{N} \to \mathbb{N}$ be the mapping

$$L(\omega,i) = \begin{cases} i+1 \text{ if } \nu\circ\theta^{-i}(\omega) > i+1, \\ 0 \text{ otherwise.} \end{cases} \qquad (4.2.4)$$

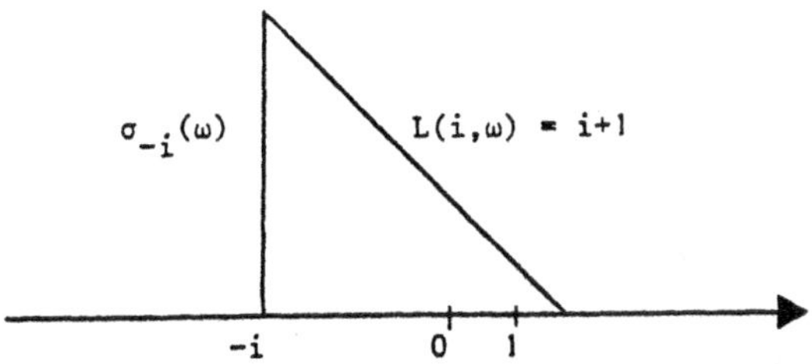

Figure 8

Let $L^m(\omega) : \mathbb{N} \to \mathbb{N}$, $m \geq 1$ be defined by

$$L^m(\omega)(i) = L(\theta^{-1}\omega) o L(\theta^{-2}\omega) o \dots o\ L(\theta^{-m}\omega)(i) \tag{4.2.5}$$

In words, $L^m(\omega,i)$ is the absolute value of the index of the customer present in the server at time 0 when assuming that customer-$(m+i)$ was accepted (see figure 9).

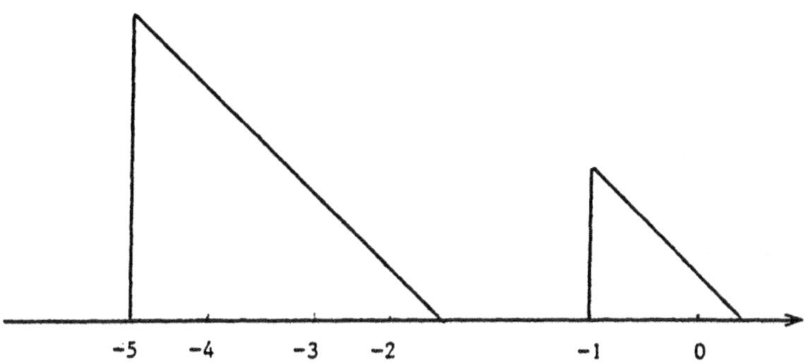

$i = 1 \qquad m = 4$

$L(\theta^{-4}(\omega),\ 1) = 2, \quad L(\theta^{-3}(\omega),\ 2) = 3, \quad L(\theta^{-2}(\omega),3) = 0, \quad L(\theta^{-1}(\omega),\ 0) = 1.$

Figure 9

Define the subset $H_m(\omega)$ of $\mathbb{N}$ by

$$H_m(\omega) = L^m(\omega,\ \mathbb{N}),\ m \geq 1. \tag{4.2.6}$$

This is the set of all the indices (up to the sign) of customers present in the server at time 0 if customer $-m-i$ was accepted, when i varies over $\mathbb{N}$. By construction $H_m(\omega) \neq \emptyset$. Using the relation

$$L^{m+1}(\omega,\mathbb{N}) = L^m(\omega) o L(\theta^{-m}(\omega)\ (\mathbb{N}),$$

one gets that $H_{m+1}(\omega) \subset H_m(\omega)$. In addition, $L(\omega,\mathbb{N}) \in [0,n]$ if $\omega \in B_n$, so that $H_m(\omega)$ is a finite set owing to (4.2.3). Therefore,

$$H_n(\omega) \downarrow H(\omega) \neq \emptyset. \qquad (4.2.7)$$

Since $L^{m+1} o\theta = L o L^m$, $H_{m+1} o\theta = L(H_m)$ and therefore card $(H_{m+1} o\theta) \leq$ card (H_m). Letting $m \uparrow \infty$, we see that card $(H o\theta) \leq$ card (H). Therefore, since (P^0, θ) is ergodic

$$\text{card } H(\omega) = \text{constant}, \ P^0 - \text{a.s.} \qquad (4.2.8)$$

We are now in a position where we can construct the enriched space $(\bar{\Omega}, \bar{F}, \bar{P}^0, \bar{\theta})$:

$$\begin{cases} \bar{\Omega} = \{(\omega,i) \in \Omega \times \mathbb{N} \ / \ i \in H(\omega)\}, \\ \bar{\mathbf{F}} = \text{trace on } \bar{\Omega} \text{ of } F_Ç \otimes \mathbf{P}(\mathbb{N}), \\ \bar{P}^0(A, \{i\}) = E^0[1_A \ \delta_{\{i\}}(H)], \\ \bar{\theta}(\omega,i) = (\theta(\omega), L(\omega,i)). \end{cases} \qquad (4.2.9)$$

$\bar{\theta}$ is an automorphism of $(\bar{\Omega}, \bar{\mathbf{F}})$. To show this, it is enough to prove that $L(\omega)$: $H(\omega) \to h(\theta(\omega))$ is bijective. But $H(\theta(\omega)) = L(\omega)(H(\omega))$ and therefore, $L(\omega)$ is surjective. It is also injective since card $(H(\omega))$ = card $H(\theta(\omega))$.

Define the mapping $f : \bar{\Omega} \to \Omega$ by $f(\omega,i) = \omega$. We have $P^0 = \bar{P}^0 o f^{-1}$ and $\theta o f^{-1} = \bar{\theta}$. Therefore, $(\bar{\Omega}, \bar{\mathbf{F}}, \bar{P}^0, \bar{\theta})$ is an enrichment of $(\Omega, \mathbf{F}, P^0, \theta)$. Notice that $\bar{P}^0$ is not necessarily a probability measure. It is however a σ-finite measure. Let us show that $\bar{P}$ is $\bar{\theta}$ invariant i.e. :

$$\int_{\bar{\Omega}} 1_{A\times\{i\}}(\omega,j) \, d\bar{P}^0(\omega,j) = \int_{\bar{\Omega}} 1_{A\times\{i\}}(\theta(\omega), L(\omega,j)) \, d\bar{P}^0(\omega,j).$$

Indeed, the right-hand side is by definition

$$\sum_{j\in\mathbb{N}} \int_{\Omega} 1_A(\theta(\omega)) \ 1_{\{i\}} \ (L(\omega,j)) \ \delta_{\{j\}}(H(\omega)) \ dP^0(\omega),$$

or, in view of the θ-invariance of P^0 :

$$\sum_{j\in\mathbb{N}} \int_{\Omega} 1_A(\Omega) \ 1_{\{i\}}(L(\theta^{-1}(\omega),j)) \ \delta_{\{j\}}(H(\theta^{-1}(\omega)) \ dP^0(\omega).$$

Since $L(\theta^{-1}(\omega))$ is a bijection from $H(\theta^{-1}(\omega))$ into $H(\omega)$, $\{i \in H(\omega)\} = \sum_{j\in\mathbb{N}} \{i = L(\theta^{-1}\omega,j), \ j \in H(\theta^{-1}\omega)\}$, which completes the proof of the $\bar{\theta}$ invariance of $\bar{P}^0$.

4.3 Construction of a stationary solution

Define $\bar{W}$ on $\bar{\Omega}$ by

$$\bar{W}(\omega,i) = \begin{cases} (\sigma o\theta^{-i} - \sum_{k=1}^{i} \tau o\theta^{-k})^+, & \text{if } i \neq 0, \\ 0 & \text{otherwise.} \end{cases} \qquad (4.3.1)$$

Since $\sigma o\theta^{-i} - \sum_{k=1}^{i} \tau o\theta^{-k} > 0$ when $i \in H(\omega)$, $i \neq 0$,

$$\overline{W}(\omega,i) = \begin{cases} \sigma o\theta^{-i} - \sum_{k=1}^{i} \tau o\theta^{-k}, \text{ if } i \neq 0, \\ 0 \text{ otherwise.} \end{cases} \tag{4.3.1)'$$

Let us check that

$$W o\overline{\theta} = (\overline{W} + \sigma 1_{\{\overline{W}=0\}} - \tau)^{+}. \tag{4.3.2}$$

<u>First case : $L(\omega,i) = i+1,\ i \neq 0$</u>

Then, $\overline{W}o\overline{\theta} = \overline{W}(\theta(\omega),\ L(\omega,i)) = \overline{W}(\theta(\omega),\ i+1) = \sigma o\theta^{-i} - \sum_{k=0}^{i} \tau o\theta^{-k}$. Since $L(\omega,i) = i+1$, $\sigma o\theta^{-i} - \sum_{k=0}^{i} \tau o\theta^{-k} > 0$ and a fortiori $\sigma o\theta^{-i} - \sum_{k=1}^{i} \tau o\theta^{-k} > 0$. i.e. $\overline{W}(\omega,i) > 0$. Hence, $\overline{W}+\sigma.1_{\{\overline{W}=0\}} - \tau = \overline{W} - \tau = \sigma o\theta^{-i} - \sum_{k=1}^{i} \tau o\theta^{-k} - \tau = \overline{W}o\overline{\theta}$.

<u>Second case : $L(\omega,i) = 0,\ i \neq 0$</u>

Then, $\overline{W}o\overline{\theta} = \overline{W}(\theta(\omega),0)$. Since $L(\omega,i) = 0$, $\sigma o\theta^{-i} - \sum_{k=0}^{i} \tau o\theta^{-k} \leq 0$. Since $i \in H(\omega)$, $\overline{W} = \sigma o\theta^{-i} - \sum_{k=1}^{i} \tau o\theta^{-k} > 0$. Therefore, $\overline{W}+\sigma.1_{\{\overline{W}=0\}} - \tau = \overline{W} - \tau = \sigma o\theta^{-i} - \sum_{k=0}^{i} \tau o\theta^{-k} \leq 0$. Thus, $(\overline{W}+\sigma.1_{\{\overline{W}=0\}} - \tau)^{+} = 0 = \overline{W}o\overline{\theta}$.

<u>Third case : $L(\omega,i) = i+1,\ i = 0$</u>

We have $L(\omega,i) = 1$ and $\overline{W}(\theta(\omega),\ L(\omega,i)) = \overline{W}(\theta(\omega),1) = (\sigma o\theta^{-1} - \tau o\theta^{-1})o\theta = \sigma - \tau$. Since $i = 0$, $\overline{W}(\omega,i) = 0$. Therefore $\overline{W} + \sigma.1_{\{\overline{W}=0\}} - \tau = \sigma-\tau$. Now $L(\omega,0) = 1$ also implies that $\sigma > \tau$, and therefore $(\overline{W}+\sigma 1_{\{\overline{W}=0\}}-\tau)^{+} = \overline{W}o\overline{\theta}$.

<u>Fourth case : $L(i,\omega) = 0,\ i = 0$</u>

Therefore $\overline{W}(\theta(\omega),\ L(\omega,i)) = \overline{W}(\theta(\omega),0) = 0$. But $L(\omega,0) = 0$, $\sigma-\tau \leq 0$ and $\overline{W}(\omega,i) = 0$, so that $\overline{W} + \sigma.1_{\{\overline{W}=0\}} - \tau = \sigma - \tau \leq 0$.

5 - OTHER QUEUEING SYSTEMS

5.1 The G/G/∞ pure delay system

The input $(\sigma_n, \tau_n, n \in \mathbb{Z})$ is defined as in §1.1. The system has an infinite reservoir of servers, so that any customer is immediately attended. Therefore, customer n arriving at time T_n leaves at time $T_n + \sigma_n$ (service is provided at unit rate).

The state of the system is aptly described by the random element $\vec{R}_n \in \overline{\mathbb{R}}^{\infty}$, where $\vec{R}_n$ is given by

$$\vec{R}_n = (R_n^1, R_n^2, R_n^3, \ldots) = \vec{R}\circ\theta^n \tag{5.1.1}$$

and $\vec{R} = (R^1, R^2, R^3, \ldots)$ is an ordering in decreasing order

$$R^1 \geq R^2 \geq \ldots \tag{5.1.2}$$

of the residual service times at time 0 :

$$\{(\sigma\circ\theta^i - |T_i|)^+ \; ; \; i \leq -1\}. \tag{5.1.3}$$

The number of non null coordinates of $\vec{R}$ is the number $X(T_0) = X(0)$ of customers in the system just before customer 0 arrives. (see figure 10).

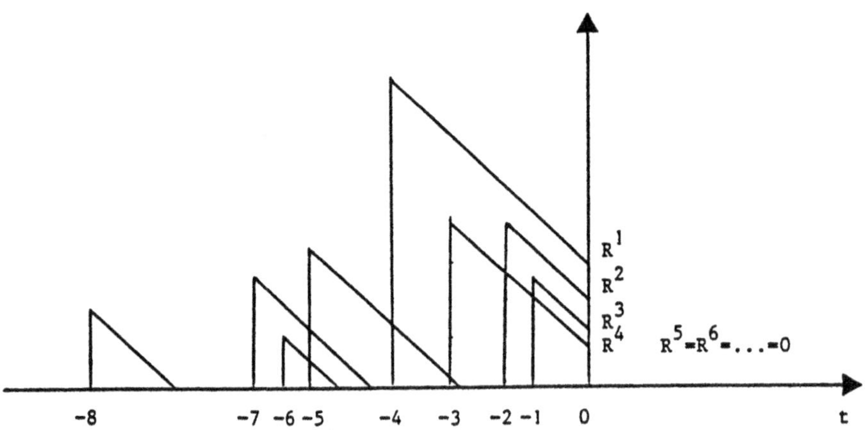

Figure 10

We will show that

$$P^0(X_0) < \infty) = 1 \tag{5.1.4}$$

and therefore, in this sense, the G/G/∞ system is "stable". It will be simpler to prove that

$$P(X(0) < \infty) = 1, \tag{5.1.5}$$

where P is the θ_t-stationary probability associated to P^0 and where X(t) is the number of customers in the system at time t. Indeed (5.1.5) implies (5.1.4) in view of the results of §8.1 of Part 1.

But $X(0) = \sum_{n\in Z/} f(T_n, \sigma_n)$ where $f(t,\sigma)$ is 1 if $x < 0$ and $\sigma > -x$, and 0 otherwise. By Campbell's formula ((3.3.2) of Part 1), and formula (3.1.4) of Part 1,

$$E[X(0)] = \int_{-\infty}^{0} \left(\int_{-t}^{+\infty} P^0(\sigma_0 \in dx)\right)\lambda dt = \lambda \int_0^{\infty} (1 - P^0(\sigma_0 \le t))dt$$

i.e. since $\lambda = 1/E^0[\sigma]$,

$$E[X(0)] = \frac{E^0[\sigma]}{E^0[\tau]} = \rho < \infty, \tag{5.1.6}$$

and therefore (5.1.5) holds. ◻

More generally, denoting $X_r(t)$ the number of busy servers at time t with a load larger or equal to $r > 0$, we get

$$E[X_r(0)] = \lambda \int_r^{\infty} (1 - P^0(\sigma \le t))dt. \tag{5.1.7}$$

Remark. The existence of an infinity of construction points (where the system is empty) to the right and to the left follows from the argument of §4.2 showing that ν defined by (4.2.1) is P^0 - a.s. finite.

5.2 The G/G/1/∞ queue in random environment

The setting is as in §1.1. and §2.1 except now that the service is not given at unit rate, but at a random rate A_t. More precisely : $(A_t, t \in \mathbb{R})$ is a non negative process, independent of $(\sigma_n, \tau_n, n \in Z\!\!/)$ (under P or equivalently under P^0), and such that $A_t o\theta_s = A_{t+s}$ $(s,t \in \mathbb{R})$ so that $(A_t, t \in \mathbb{R})$ is a P-stationary process. The maximum amount of service which can be provided during the inverval $[T_n, T_{n+1}]$ is

$$\tilde{\tau}_n = \int_{T_n}^{T_{n+1}} A_s\, ds \tag{5.2.1}$$

and $\tilde{\tau}_n = \tilde{\tau} o\theta^n$ $(n \in Z\!\!/)$ on $\{T_0 = 0\}$. The evolution equation for the load process is $\tilde{W}_{n+1} = (\tilde{W}_n + \sigma_n - \tilde{\tau}_n)^+$.

As we saw in §1, a sufficient condition for the existence of a finite stationary load $\tilde{W}$ is $E^0[\sigma] < E^0[\tilde{\tau}]$. Also $E^0[\sigma] > E^0[\tilde{\tau}]$ implies that there exists no such finite stationary load. But now

$$E^0[\tilde{\tau}] = E^0[\int_0^{T_1} A_s ds] = E^0[\int_0^{T_1} A_0 o\theta_s\, ds]$$

and therefore, by the inversion formula (4.1.2a) of Part 1, $E^0[\int_0^{T_1} A_0 o\theta_s ds] = \frac{1}{\lambda} E[A_0]$, where $\frac{1}{\lambda} = E^0[\tau]$. Thus

$$E^0[\tilde{\tau}] = E[A_0] \,.\, E^0[\tau]. \tag{5.2.2}$$

Therefore if

$$E^0[\sigma] < E^0[\tau] \,.\, E[A_0] \tag{5.2.3}$$

there exists a finite stationary workload $\tilde{W} = \tilde{M}_\infty$ where $\tilde{M}_\infty = \lim_{n\to\infty} \uparrow \tilde{M}_n$ and

$$\begin{cases} \tilde{M}_0 = 0 \\ \tilde{M}_{n+1} o\theta = (\tilde{M}_n + \sigma - \tilde{\tau})^+. \end{cases} \tag{5.2.4}$$

We wish to compare the moments of $\tilde{W}$ to those of W, the stationary workload corresponding to a uniform service rate $a = E[A_0]$: We have $W = M_\infty = \lim_{n\to\infty} \uparrow M_n$ where M_n is given by

$$\begin{cases} M_0 = 0 \\ M_{n+1} o\theta = (M_n + \sigma - a.\tau)^+. \end{cases} \tag{5.2.5}$$

Notice that W will also be finite under condition (5.2.3). If (5.2.3) is satisfied, we have

(5.2.6) $E^0[f(\tilde{W})] \geq E[f(W)]$ <u>for all convex increasing function $f : \mathbb{R}^+ \to \mathbb{R}^+$</u>.

To show this, call G the σ-field generated by $(\sigma_n, \tau_n, n \in \mathbb{Z})$. We will show by induction that

$$E^0[\tilde{M}_n \mid G] \geq M_n, \; n \geq 0. \tag{5.2.7}$$

The induction hypothesis is true for $n = 0$.

Suppose it is true for n. By Jensen's inequality

$$E^0[(\tilde{M}_n + \sigma - \tilde{\tau})^+ \mid G] \geq (E^0[M_n \mid G] + \sigma - E^0[\tilde{\tau} \mid G])^+. \tag{5.2.8}$$

We have also

$$E^0[\tilde{\tau} \mid G] = E^0[\int_0^{T_1} A_s ds \mid G] = \int_{-\infty}^{+\infty} E^0[A_s 1_{[0,T_1]} \mid G]ds.$$

Owing to the independence assumption this reads

$$\int_0^{T_1} E^0[A_s \mid G]ds = \int_0^{T_1} E^0[A_s]ds = a\tau,$$

where we have also used the fact that $P = P^0$ on $\sigma\{(A_t, t \in \mathbb{R})\}$. The induction assumption gives $E^0[\tilde{M}_n \mid G] \geq M_n$. Therefore, from (5.2.8)

$$E^0[\tilde{M}_{n+1} \circ \theta \mid G] \geq (M_n + \sigma - \sigma\tau)^+ = M_{n+1} \circ \theta.$$

If we can show that

$$E^0[\tilde{M}_{n+1} \circ \theta \mid G] = E^0[\tilde{M}_{n+1} \mid G] \circ \theta, \tag{5.2.9}$$

the induction will be established. To show this, take A in G. Then,

$$E^0[1_A E^0[\tilde{M}_{n+1} \circ \theta \mid G] = E^0[1_A \tilde{M}_{n+1} \circ \theta] = E^0[1_{\theta A} \tilde{M}_{n+1}] = E^0[1_{\theta A} E^0[\tilde{M}_{n+1} \mid G]] =$$

$$= E^0[1_A E^0[\tilde{M}_{n+1} \mid G] \circ \theta],$$

since $A \in G \Rightarrow \theta A \in G$.

From (5.2.7) and Jensen's inequality, we get that for f as in (5.2.6),

$$E^0[f(\tilde{M}_n) \mid G] \geq f(M_n), \; n \geq 0.$$

The proof of (5.2.6) is then concluded when letting n go to infinity. □

<u>Remark.</u> <u>An extremal property of D/G/1/∞ queues.</u>

Consider a G/G/1/∞ queue where the service requirements are independent of the interarrival times. One proves by the same arguments that the queue $(\tilde{\tau}, \sigma)$ with interarrival times $\tilde{\tau} = E^0[\tau]$ is always "smaller" than the original queue in the sense defined by (5.2.6). A symmetrical result hods for G/D/1/∞ queues.

5.3 Priorities in G/G/1/∞ : the vector of residual service times

We are now going to introduce service disciplines in the general G/G/1/∞ model of §1. The stability condition $E^0[\sigma] < E^0[\tau]$ is supposed to hold, and therefore there exists a unique finite stationary load W.

We shall construct a state $\vec{S}$ richer in information than W, which features in particular the service discipline (LIFO, FIFO, SPRT, ...).

$\vec{S}$ is an infinite vector of non negative numbers, i.e. $\vec{S} \in \mathbb{R}_+^{\mathbb{N}}$, and only the I first coordinates of $\vec{S}$ are non null, where I is random. $\vec{S}$ is the vector of residual <u>service times</u> of the I customers present in the system <u>just before time 0.</u> In position I, one finds the customer being presently served, in position I-1 the next customer to be served, etc...

Customer 0 arrives, and this possibly changes the priorities. The priority vector at time +0 is $\vec{S}^+$:

$$\vec{S}^+(\omega) = \Phi(\omega)\,(\vec{S}(\omega) + \sigma(\omega)\,\vec{e}_{I(\omega)+1}), \tag{5.3.1}$$

where $\vec{e}_i$ is the ith vector of the canonical base of $\mathbb{R}^{\mathbb{N}}$, and $\Phi(\omega)$ is a permutation on $\mathbb{R}^{\mathbb{N}}$ involving only the $I(\omega)+1$ first coordinates. Φ describes service disciplines.

EXAMPLES

FIRST-IN-FIRST-OUT (FIFO) : (5.3.2)

$\Phi = (I+1, 1, 2, \dots, I) \quad (I \geq 1).$

LAST-IN-FIRST-OUT (LIFO) without preemption (5.3.3)

$\Phi = (1, 2, \dots, I-1, I+1, I).$

There is no preemption I.e. the last customer arriving does not interrupt the service of customer being serviced.

LAST-IN-FIRST-OUT with preemption (5.3.4)

$\Phi = (1, 2, \dots, I, I+1) = \text{identity}.$

SHORTEST-REMAINING-PROCESSING-TIME (SPRT). It is a priority discipline with preemption. Denoting

$\vec{S} = (S^1, S^2, \dots, S^I, 0, \dots)$

we define SPRT by :

$$\Phi = (I+1, 1, 2, \dots, I), \text{ if } S^1 \leq \sigma, \tag{5.3.5}$$

$\Phi = \text{identity, if } S^I > \sigma,$

$\Phi = (1, 2, \dots, j-1, I+1, j, \dots, I) \text{ if } S^j \leq \sigma < S^{j-1} \quad (2 \leq j \leq I).$

SHORTEST-PROCESSING-TIME (SPT). It is the same discipline as SPRT, only without preemption. The only change with respect to SPRT is in the case $S^I > \sigma$ where, for SPT, $\Phi = (1, 2, \dots, I-1, I+1, I)$.

In the above disciplines, the arrival of a new customer creates a modification of the order of service which depends only upon $\vec{S}$ and σ. There exist more complex disciplines for which Φ depends upon the whole sequence $(\sigma\circ\theta^{-n}, \vec{S}\circ\theta^{-n} ;\ n \geq 0)$. We shall consider here only these disciplines which depend upon

$$(\sigma\circ\theta^{-n}, \vec{S}\circ\theta^{-n} ;\ 0 \leq n \leq \nu), \tag{5.3.6}$$

where

$$\nu = \inf \{n \in \mathbb{N} \mid W\circ\theta^{-n} = 0\}. \tag{5.3.9}$$

Such disciplines will be said to be admissible. For each admissible discipline Φ, there exists a corresponding unique stationary state $S = S_\Phi$. The actual construction uses the construction points has follows :

$$\vec{S}_{\Phi,n} = 0, \text{ if } n \text{ is a construction point } (W\circ\theta^{-n} = 0), \tag{5.3.10a}$$

$$\vec{S}^{+}_{\Phi,n} = \Phi(\vec{S}_{\Phi,n} + \sigma_n \vec{e}_{I_n+1}), \tag{5.3.10b}$$

$$S^{j}_{\Phi,n+1} = (\sum_{k=j}^{\infty} (\vec{S}^{+}_{\Phi,n})^k - \tau_n)^+ - (\sum_{k=j+1}^{\infty} (\vec{S}^{+}_{\Phi,n})^k - \tau_n)^+. \tag{5.3.10c}$$

The last equality (5.3.10c) expresses that the residual service times are consumed at unity rate, in the order of the customers (see figure below).

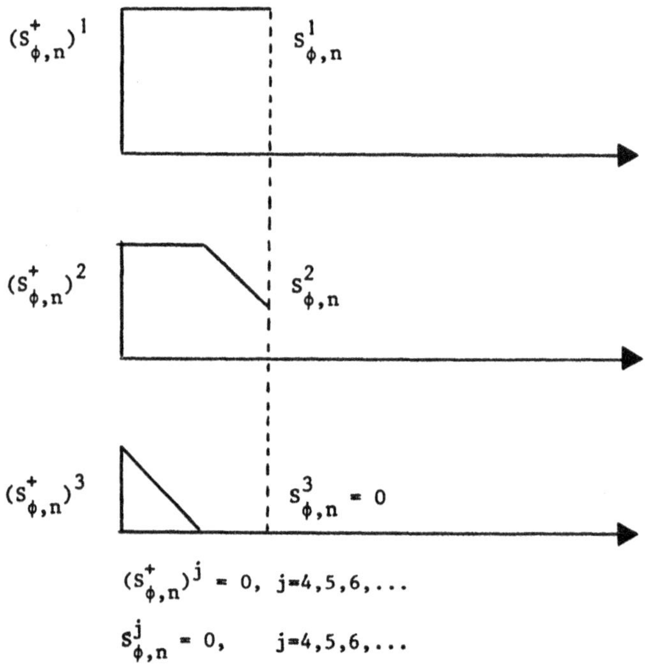

Figure 11

Construction of the time-stationary states

Let now P be the stationary probability associated with P^0. Let Φ be an admissible priority discipline. From the sequence of marks $(S_{\Phi,n}, n \in \mathbb{Z})$ and $(S^{+}_{\Phi,n}, n \in \mathbb{Z})$ one can construct the process of residual service times $(S_{\Phi}(t), t \in \mathbb{R})$ by

$$\vec{S}_{\Phi}(T_n) = \vec{S}^{+}_{\Phi,n}, \; (n \in \mathbb{Z}), \tag{5.3.11a}$$

$$S^{j}_{\Phi}(t) = (\sum_{k=j}^{\infty} (\vec{S}^{+}_{\Phi,n})^k - t)^+ - (\sum_{k=j+1}^{\infty} (\vec{S}^{+}_{\Phi,n})^k - t)^+,$$

$$(t \in [T_n, T_{n+1}) \; ; \; n \in \mathbb{Z}). \tag{5.3.11b}$$

This construction must be shown to be P-consistent, i.e. one must show that the evolution dynamics are respected, i.e.

$$\vec{S}_\Phi(T_n) = \Phi(\vec{S}_\Phi(T_n^-) + \sigma_n \vec{e}_{I_n+1}), \ (n \in Z/)\ P - \text{a.s.} \tag{5.3.12}$$

But $\vec{S}_\Phi(T_n^-) = \vec{S}_{\Phi,n}$ in view of (5.3.11b) and (5.3.10c), and $\vec{S}_\Phi(T_n) = \vec{S}^{+}_{\Phi,n}$ by definition (5.3.11a) and therefore (5.3.12) reduces to

$$\vec{S}^{+}_{\Phi,n} = \Phi(\vec{S}_{\Phi,n} + \sigma_n \vec{e}_{I_n+1}), \ (n \in Z/).$$

But this being P^0 - a.s. true, (5.3.12) is verified in view of the invariance results of §8.1, Part 1.

5.4 Optimality properties of the SPRT rule

(5.4.1) <u>If Ψ denotes the SPRT discipline, then for any other admissible discipline $\vec{S}_\psi < \vec{S}_\Phi$ in the sense that</u>

$$\sum_{i=k}^{\infty} S_\Phi^i \geq \sum_{i=k}^{\infty} S_\Psi^i, \ k \geq 0.$$

<u>Proof.</u> Let $\tilde{\vec{S}}$ be a reordering of $\vec{S} \in \mathbb{R}^{\mathbb{N}}$ in decreasing order, i.e. $\tilde{S}^1 \geq \tilde{S}^2 \geq \ldots$ Since $\tilde{\vec{S}}_\Psi = \vec{S}_\Psi$ and $\tilde{\vec{S}}_\Phi < \vec{S}_\Phi$, it suffices to show that $\vec{S}_\Psi < \vec{S}_\Phi$.

Since $\tilde{\vec{S}}_\psi = \vec{S}_\psi = \vec{S}_\Phi = \tilde{\vec{S}}_\Phi = \vec{0}$ on $W = 0$, and $P^0(W = 0) > 0$, if we can show that $\{\tilde{\vec{S}}_\psi < \tilde{\vec{S}}_\Phi\}$ is a θ-invariant event, then it has necessarily probability one, which we seek to prove.

Since

$$\sum_{k=j}^{\infty} S_\Phi^i \circ\theta = (\sum_{k=j}^{\infty} (\vec{S}^{+}_\Phi)^k - \tau)^+,$$

we have for $j > 1$:

$$\sum_{k=j}^{\infty} \tilde{S}_\psi^k \circ\theta = (\sum_{k=j}^{\infty} \tilde{S}_\Psi^k + \sigma - \tau)^+ \wedge (\sum_{k=j-1}^{\infty} \tilde{S}_\Psi^k - \tau)^+$$

$$\sum_{k=j}^{\infty} \tilde{S}_\Phi^k \circ\theta \geq (\sum_{k=j}^{\infty} \tilde{S}_\Phi^k + \sigma - \tau)^+ \geq (\sum_{k=j-1}^{\infty} \tilde{S}_\Phi^k - \tau)^+$$

and for $j = 1$:

$$\sum_{k=1}^{\infty} \tilde{S}_\Phi^k = \sum_{k=1}^{\infty} \tilde{S}_\Psi^k = W.$$

Therefore $\{\vec{\tilde{S}}_\psi o\theta < \vec{\tilde{S}}_\Phi o\theta\} \supset \{\vec{\tilde{S}}_\psi < \vec{\tilde{S}}_\Phi\}$, which implies the invariance of $\{\vec{\tilde{S}}_\psi < \vec{\tilde{S}}_\Phi\}$ since P is θ-invariant. □

Since $\{\vec{S}_{\Psi,n} < \vec{S}_{\Phi,n}\ (n \in Z/)\} \equiv \{\vec{S}_\Psi(t) < \vec{S}_\Phi(t)\ (t \in \mathbb{R})\}$ another application of the invariance results of §8.1, Part 1, yields, in view of (5.4.1):

(5.4.2) <u>If Ψ is the SPRT discipline and Φ is any other discipline</u>

$$\vec{S}_\Psi(t) < \vec{S}_\Phi(t)\ (t \in \mathbb{R}),\ P-\text{a.s.}$$

in particular:

$$X_\Psi(t) < X_\Phi(t)\ (t \in \mathbb{R}),\ P-\text{a.s.} \tag{5.4.3}$$

where $(X_\Phi(t),\ t \in \mathbb{R})$ is the process counting the number of customers present in the system at time t under discipline Φ.

Application of Little's formula shows that

$$E^0[W_\Psi] \le E^0[W_\Phi], \tag{5.4.4}$$

for all admissible disciplines Φ.

6 THE BEDIENUNGSSYSTEME

6.1 - The mechanism and the input

Let G be a denumerable set. An element $g \in G$ is called a <u>macrostate</u>.

Let S be another denumerable set. An element $s \in S$ is called a <u>site</u>.

With each macrostate $g \in G$ are associated two finite subsets of S, P(g) and A(g), respectively called the set of sites <u>present</u> in g, and the set of site <u>active</u> in g, and such that

$$A(g) \subset P(g) \subset S \tag{6.1.1}$$

With each pair (s,g) such that $s \in P(g)$ is associated a number c(s,g) called the <u>speed of consumption</u> on site s, in state g. The speeds c(s,g) verify the conditions

$$\begin{cases} c(s,g) > 0 & \text{if } s \in A(g), \\ c(s,g) = 0 & \text{if } s \notin A(g), \\ \sum_{s\in A(g)} c(s,g) > 0. \end{cases} \tag{6.1.2}$$

In particular for all $g \in G$, the set of active sites A(g) is not empty.

With each pair $(s,g) \in S \times G$ is associated a measure $p(g,s,.)$ on G such that :

$$\begin{cases} \sum_{g' \in G} p(g,s,g') = 1 & \text{if } s \in A(g), \\ p(g,s,g') = 0 \quad (g' \in G) & \text{if } s \notin A(g). \end{cases} \tag{6.1.3}$$

If $p(g,s,g') > 0$, one says that transition $g \overset{s}{\to} g'$ is allowed. In this case, it is assumed that

$$P(g) - \{s\} \subset P(g') \tag{6.1.4}$$

i.e. all sites of g, except may be s, remain present in the new state g' of transition $g \overset{s}{\to} g'$. In the sequel, we shall also use the notation G_s to represent the set of macrostates in which s is present : $G_s = \{g \in G / s \in P(g)\}$.

The quadruple

$$\sum = \{G, S, c(.,.), p(.,.,.)\} \tag{6.1.5}$$

is called a <u>mechanism</u>.

With each site $s \in S$ is associated a point process in canonical form $(M_s, \mathbf{M}_s, \mathbf{P}_s)$, which is stationary (i.e. $\mathbf{P}_s \circ T_t^s = \mathbf{P}_s$ $(t \in \mathbb{R})$, where $(T_t^s, t \in \mathbb{R})$ is the semi-group of shifts acting on M_s). It will be assumed moreover that the intensity λ_s of $(M_s, \mathbf{M}_s, \mathbf{P}_s)$ verifies

$$0 < \lambda_s < \infty \quad (s \in S). \tag{6.1.6}$$

$\mathbf{P}_s^0$ denotes the Palm probability associated with $\mathbf{P}_s$. The generic element of M_s is denoted by m_s, and the identy mapping from M_s into M_s by Φ_s. Therefore, $(\Phi_s, T_t^s, \mathbf{P}_s)$ is a stationary point process.

The family $\{(M_s, \mathbf{M}_s, \mathbf{P}_s), s \in S\}$ is called the <u>input</u> of the mechanism.

We shall use the following notations

$$\begin{aligned} \lambda(s,g) &= \lambda_s \, c(s,g), \\ \Lambda(g) &= \sum_{s \in A(g)} \lambda_s \, c(s,g). \end{aligned} \tag{6.1.7}$$

6.2 - A heuristic description of the dynamics

Suppose that to each site $s \in S$ is attached a trajectory $m_s \in M_s$. We shall denote as $(t_n^s, n \in \mathbb{Z})$ the sequence of points of m_s. As a matter of fact only the restriction of m_s to $\mathbb{R}_+$ is needed, i.e. $(t_n^s, n \in \mathbb{N})$.

Each site $s \in S$ has a little insect B_s whose single occupation is to "gnaw" m_s. At each instant t, the gnawing speed of B_s is $c(s,g)$ if the "system" is in state g. If we call $X(t)$ the <u>state</u> of the system at time t, then the beast B_s will be at position $\theta_s(t) = \int_0^t c(s,X(u))du$ at time t (it starts from 0 at time 0).

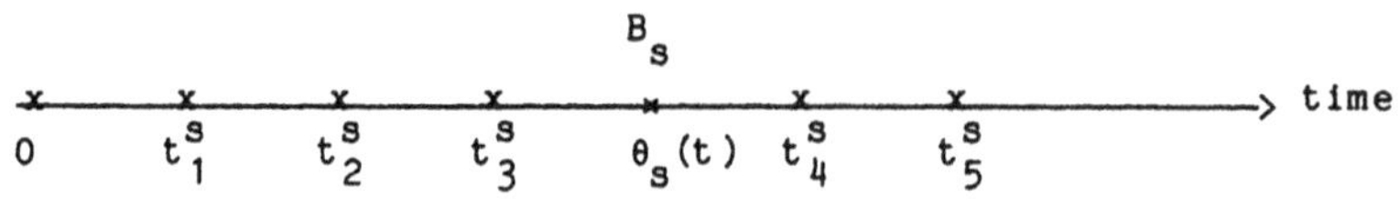

The speed $c(s,X(t))$ at time t is non-null if and only if s is an active site of state $X(t) \in G$.

When a beast B_s reaches a point t_n^s of the trajectory m_s, an "event" is said to occur. The state of the system changes. If the state is g, and if the "event" is due to $s \in A(g)$ (i.e. B_s reaches a point), the new state g' is chosen according to the probability distribution $p(g,s,.)$. One then says that the transition $g \to g'$ has been triggered on site s. We recall the notation $g \overset{s}{\to} g'$ for this.

We shall see in the formal construction of the evolution process of the system, that under the assumptions of the model, there cannot be simultaneous "events", i.e. that two different beasts will not reach simultaneously a point of their point process trajectory.

At each time t, the generalized state of the system has the form $\hat{Z}(t)=(X(t),\hat{Y}(t))$ where $\hat{Y}(t)$ represents what is left of the trajectories m_s attached to each site $s \in S$. More precisely

$$\hat{Y}(t) \in \prod_{s\in S} \hat{M}_s, \quad \hat{Y}(t) = (\widehat{T^s_{\theta_s(t)}\, m_s}, \ s \in S), \tag{6.2.1}$$

where

$$\theta_s(t) = \int_0^t c(s,X(u))du \tag{6.2.2}$$

and where the $\widehat{\ }$ represents the following operation on the trajectories

$$\hat{m}_s = \begin{cases} \text{restriction of } m_s \text{ to } [0,\infty) \text{ if } s \in S', \\ \\ \text{restriction of } m_s \text{ to } (0,\infty) \text{ if } s \in S-S', \end{cases} \tag{6.2.3}$$

for some subset $S' \subset S$ to be precised later.

For more precision in the notations, we will replace in (6.2.1) m_s by Φ_s, just to emphasize the fact that $(m_s, s \in S)$ is chosen randomly :

$$Y(t) = (T^s_{\theta_s(t)}\, \Phi_s, \ s \in S), \quad Z(t) = (X(t), Y(t)), \quad \hat{Z}(t) = (X(t), \hat{Y}(t) \tag{6.2.4}$$

6.3 - The initial generalized state

The initial generalized state of the system is defined as follows. Let $(\Omega, \mathbf{F}, P^0)$ be the canonical probability space of $(Z(t), t \geq 0)$. Denote

$$S = \{s_1, s_2, \dots\}, \quad A(g) = \{s_{i_1}, \dots s_{i_{|A(g)|}}\}. \tag{6.3.1}$$

For any subet $A \subset S$ let

$$P_A = \bigotimes_{i=1}^{\infty} Q_{s_i,A} \tag{6.3.2}$$

where

$$Q_{s,A} = \begin{cases} P_s & \text{if } s \in A, \\ P_s^0 & \text{if } s \notin A. \end{cases} \tag{6.3.3}$$

Define, for each $g \in G$, a probability measure $\hat{P}_{A(g)}$ on $(\prod_{s\in S} M_s, \bigotimes_{s\in S} \mathcal{M}_s)$ by

$$\hat{P}_{A(g)} = \sum_{\ell=1}^{|A(g)|} \frac{\lambda(s_{i_\ell},g)}{\Lambda(g)} \left(\bigotimes_{i=1}^{i_\ell-1} Q_{s_i,A(g)} \right) \otimes P^0_{s_{i_\ell}} \otimes \left(\bigotimes_{i=i_\ell+1}^{|A(g)|} Q_{s_i,A(g)} \right). \tag{6.3.4}$$

It will be noticed for future reference that

$$\hat{P}_{A(g)} = \sum_{\ell=1}^{|A(g)|} \frac{\lambda(s_{i_\ell},g)}{\Lambda(g)} P_{A(g)-\{s_{i_\ell}\}}. \tag{6.3.5}$$

Let now $(x_g^0, g \in G)$ be a probability distribution to be precised later.

The initial generalized state is chosen according to

$$P^0(X(0) = g, Y(0) \in B) = x_g^0 \hat{P}_{A(g)}(B) \quad (g \in G, B \in \bigotimes_{s\in S} \mathcal{M}_s). \tag{6.3.6}$$

The interpretation of (6.3.6) is as follows. First select an initial state $X(0)$ according to the probability distribution $(x_g^0, g \in G)$. Once $X(0) = g$ is selected, select a site $\sigma_0 \in S$ according to the probability distribution $\left(\frac{\lambda(s,g)}{\Lambda(g)}, s \in S \right)$.

Note that since $\lambda(s,g) = 0$ when $s \notin A(g)$, σ_0 is necessarily an active site of $X(0) = g$.

Once $X(0) = g$ and $\sigma_0 = s \in A(X(0))$ have been selected, select independently the trajectories $\Phi_{s_i} \in M_{s_i}$ $(s_i \in S)$ according to P_{s_i} if $s_i \in A(g) - \{s\}$, and according to $P^0_{s_i}$ if $s_i \notin A(g)-\{s\}$. This determines $Y(0) = (\Phi_{s_i}, s_i \in S)$.

6.4 - The evolution

Let X_1^s be the first point of Φ_s strictly to the right of the origin of times and define

$$T_1 = \inf\left\{ \frac{X_1^s}{c(s,X(0))}, s \in A(X(0)) \right\} \tag{6.4.1}$$

and let $\tilde{\sigma}_0$ be the site of $A(X(0))$ realizing the inf in (6.4.1) (We shall see that $\tilde{\sigma}_0$ is unambiguously defined).

In the terms of the heuristic description of § 6.2, the beast $B_{\tilde{\sigma}_0}$ has reached a point of its attached point process $\Phi_{\tilde{\sigma}_0}$ before the other insects in competition.

(6.4.2) $\tilde{\sigma}_0$ is well defined.

Proof : The point processes $(\Phi_{s_i}, s_i \in S)$ forming Y(0) are independent given X(0) = g and σ_0 = s. Moreover in A(X(0)) = A(g), only one site contributes to the expression of $P_{A(g)-\{s\}}$ with a Palm probability. Therefore, given X(0) = g and σ_0 = s, the random variables $\frac{X_1^{s'}}{c(s',g)}$ (s' $\in$ A(g)) take P^0-a.s. different values since they are independent and they all admit -except maybe one- probability densities. □

Clearly T_1 is almost surely positive.

T_1 is the first "event". It generates a change of state. The state X(t) remaines the same (=X(0)) in the time interval $[0,T_1)$. It then changes from $X(T_1-)=X(0)=g$ to $X(T_1)$. The probability distribution of $X(T_1)$ is conditionally independent of Y(0) given X(0) and $\tilde{\sigma}_0$ and its conditional probability distribution is $p(X(0),\tilde{\sigma}_0,\cdot)$. Site $\tilde{\sigma}_0$ triggers the transition, i.e.

$$X(T_1-) = X(0) \xrightarrow{\tilde{\sigma}_0} X(T_1).$$

For all t ≥ 0, Y(t) is defined by (6.2.4). We shall see that, under certain assumptions to be precised in § 8, if we start with an initial distribution of Z(0) = (X(0),Y(0)) given by (6.3.6), then $(X(T_1), Y(T_1))$ has the same distribution as (X(0),Y(0)). We continue to define the generalized state Z(t) on t ≥ T_1, using the same procedure as above, but starting now with the new state $X(T_1)$ and with what remains of the initial point processes.

This gives rise to a sequence of events

$$T_0 = 0 < T_1 < T_2 < \dots \qquad (6.4.3)$$

and also to sequences $(\sigma_n, n \in \mathbb{Z})$, $(\tilde{\sigma}_n, n \in \mathbb{Z})$ where σ_n and $\tilde{\sigma}_n$ are defined analogously to σ_0 and $\tilde{\sigma}_0$ respectively. In particular $\tilde{\sigma}_n$ triggers the transition at time T_{n+1}

$$X(T_{n+1}^-) = X(T_n) \xrightarrow{\tilde{\sigma}_n} X(T_{n+1}) \quad (n \in \mathbb{N}). \qquad (6.4.4)$$

This procedure defines Z(t) = (X(t),Y(t)) only on $[0,T_\infty)$ where

$$T_\infty = \lim T_n. \qquad (6.4.5)$$

Define

$$X_n = X(T_n). \qquad (6.4.6)$$

(6.4.7) We shall say that the Bedienungssysteme is regular if $T_\infty = \infty$, P^0-a.s.

6.5 Examples

(6.5.1) The queue G/G/1/∞/LIFO preemptive.

This is a waiting system with one station, one server, and an infinite capacity for the waiting room. The customers are served in the inverse order of arrival, and the customer just arriving receives immediate attention from the server who leaves the customer he was serving and sends him back to the waiting room.

A mechanism corresponding to the above description is the following one

$$\begin{cases} S = \mathbb{N}, \\ G = \{(0,1,\ldots,n)\ ,\ n \in \mathbb{N}\}\ ;\ (g_n = (0,1,\ldots,n)), \\ P(g_n) = \{0,1,\ldots,n\}, \\ A(g_n) = \{0,n\}. \end{cases} \tag{6.5.1}$$

The only non zero speeds are

$$c(o,g_n) = c(n,g_n) = 1. \tag{6.5.2}$$

Interpretation : 0 is the site of the source, which is always active (it is an infinite reservoir of customers, and all customers are accepted in the system). When the system is in macrostate g_n, there are n customers in the positions (sites) 1,...,n. The customer in site n is served at unit speed. The other ones wait. The positions of the customers indicate their order of arrival. The only transitions allowed are

$$\begin{cases} g_n \overset{o}{\to} g_{n+1}, & \text{(arrival)} \\ g_{n+1} \overset{n}{\to} g_n & \text{(departure)} \end{cases} \tag{6.5.3}$$

This leaves no choice for $p(.,.,.)$.

If for the site o, (M_o,M_o,P_o) is a Poisson process with intensity λ, and if for all other sites $n \geq 1$, (M_n,M_n,P_n) is a Poisson process with intensity μ, the process $(X(t),\ t \geq o)$ counting (with the indentification $g_n \equiv n$) the number of customers in the system is a Markov process with state space $\mathbb{N}$ and transition matrix $Q = \{q_{n,n'}\ ;\ n,n' \in \mathbb{N}\}$ defined by

$$q_{n,n+1} = \lambda\ ,\ q_{n+1,n} = \mu \tag{6.5.4}$$

In other words $(X(t),\ t \geq 0)$ is an M/M/1/∞ process.

6.5.2 The Erlang system G/G/k/o

It is a waiting system with one station, k servers, and no waiting room : a customer who finds the k servers busy is lost.

This system can be described by the following mechanism

$$\begin{cases} S = \{o,1,\dots,k\}, \\ G = \{\text{all subsets of } S \text{ containing } 0\}, \\ P(g) = g. \end{cases} \tag{6.5.5}$$

All sites present in a macrostate are active with unit speed :

$$\begin{cases} A(g) = g, \\ c(s,g) = 1 \ (s \in g). \end{cases} \tag{6.5.6}$$

Site 0 is the source site. All other sites are equipped with one server. Two macrostates are of particular importance : $g_o = \{o\}$, $g_k = S = \{o,1,\dots,k\}$. The allowed transitions are :

$$\begin{cases} g \neq g_o : g \overset{i}{\to} g-\{i\} \quad (i \in g,\ i \neq o) & \text{server } i \text{ releases a customer.} \\ g \neq g_k : g \overset{o}{\to} g+\{j\} \quad (j \notin g,\ j \neq o) & \text{the source creates a customer who is assigned to server } j. \\ \qquad g_k \overset{o}{\to} g_k & \text{The source creates a customer who is rejected, the system being full.} \end{cases} \tag{6.5.7}$$

In the transition $g \overset{o}{\to} g+\{j\}$, the server j is chosen according to the probability distribution $(\pi(g,j)\ ;\ j \in (S-g)$, i.e.

$$p(g,o,g+\{j\}) = \pi(g,j). \tag{6.5.8}$$

The intensities associated to the sites are

$$\begin{cases} \lambda_o = \lambda, \\ \lambda_i = \mu_i \quad (1 \le i \le k). \end{cases} \tag{6.5.9}$$

(6.5.10) <u>For virtually all queueing systems, the corresponding Bedienungssystem is regular.</u>

The case of a G/G/1/∞ model contains all the basic features which make things work in most queueing models of interest, such as BCMP networks for instance.

The arrival process (and every process (M_s,M_s,P_s) associated with a site in the general model) is regular in the sense that its points satisfy the condition (6.4.7). It is therefore sufficient, in order to show the regularity of a queueing model, to prove that there are only a finite number of events (changes of state) between two consecutive arrivals τ_j and τ_{j+1}.

Suppose that after the first arrival there are n customers in the system. Then, before the next arrivals, only the states $g_0,\dots,g_n$ can be visited, and only the sites $1,\dots,n$ can be active. Therefore the supremum c^* of the speeds $c(s,g)$ that occur between two arrivals is finite. The number of events in (τ_j,τ_{j+1}) is then obviously bounded by $\sum_{k=1}^{n} m_{s_k}([\tau_j,\tau_j+c^*(\tau_{j+1} - \tau_j)])$ where m_{s_k} is the point process trajectory which remains to be consumed at site s_k and at time τ_j. This number is finite because the input point processes are regular and c^* is finite.

7 THE INSENSITIVITY BALANCE EQUATIONS

7.1 - Stability and regularity assumptions

Consider the case where all the point processes (M_s, M_s, P_s) $(s \in S)$ are Poisson. It is clear that $(X_n, n \geq 0)$ is a Markov chain with state space G and transition matrix $\mathbb{P} = \{p_{g,g'}\ ;\ g, g' \in G\}$ defined by

$$p_{g,g'} = \Big(\sum_{s \in A(g)} \lambda(s,g)\ p(g,s,g') \Big) / \Lambda(g). \tag{7.1.1}$$

One makes the following assumption :

A1 : The matrix $\mathbb{P}$ defined by (7.1.1) is ergodic.

In particular, there exists a unique stationary distribution $(x_g^o, g \in G)$ given by

$$x_g^o = \sum_{g' \in G} x_{g'}^o \Big(\sum_{s' \in A(g')} \lambda(s',g')\ p(g',s',g) \Big) / \Lambda(g'). \tag{7.1.2}$$

The second assumption made on the system is :

A2 : The stationary distribution $(x_g^o, g \in G)$ defined by (7.1.2) verifies :

$$\sum_{g \in G} (x_g^o / \Lambda(g)) < \infty. \tag{7.1.3}$$

In particular, there exists a normalizing constant K, $o < K < \infty$, such that the measure on $G(x_g, g \in G)$ defined by

$$x_g^o = K\ x_g\ \Lambda(g) \tag{7.1.4}$$

is a probability measure and verifies

$$x_g\ \Lambda(g) = \sum_{g' \in G} x_{g'} \sum_{s' \in A(g')} p(g',s',g)\ \lambda(s',g'). \tag{7.1.5}$$

If there exists a regular continuous time Markov chain with state space G and an infinitesimal generator given by

$$q_{g,g'} = \sum_{s \in A(g)} p(g,s,g')\ \lambda(s,g), \tag{7.1.6}$$

then $(x_g, g \in G)$ is a stationary distribution for it and the imbedded Markov chain admits $\mathbb{P}$ for transition matrix.

As a matter of fact, if (M_s, M_s, P_s) is a Poisson processes for all $s \in S$, $(X(t), t \geq 0)$ is a continuous time Markov chain with the generator given by (7.1.6). The result stating that $T_\infty = \infty$ (regularity) has to be verified case by case. A sufficient condition for regularity is that $\sup_{g \in G} \Lambda(g) < \infty$. However these exist systems for which we have regularity and such that $\sup \Lambda(g) = \infty$ like the M/M/∞ queue. But we have seen that regularity holds for virtually all queueing systems (cf. (6.5.10)).

Equation (7.1.5) will be referred to as the <u>global balance</u> equation.

7.2 – Insensitivity balance equations

Recall the existence of a subset $S' \subset S$, called the set of insensitive sites. The definition of S', or, equivalently, the third assumption on the mechanism is the following.

A3 : For all $s_0 \in S'$ and all $g \in G_{s_0}$ (i.e. $s_0 \in P(g)$),

$$x_g \lambda(s_0,g) = \sum_{g' \notin G_{s_0}} x_{g'} \sum_{s \in A(g')} p(g',s,g)\lambda(s,g') + \sum_{g' \in G_{s_0}} x_{g'} p(g',s_0,g) \lambda(s_0,g') \tag{7.2.1}$$

Equation (7.2.1) is called the insentivity balance equation relative to s_0 and $g \in G_{s_0}$.

Interpretation : if $g \overset{s}{\to} g'$ and $s \in A(g')$, one says that a pseudo-death of s (followed by its pseudo-birth) has occured. Otherwise, that is to say if $s \notin A(g')$, we have an actual death of s. If $g \overset{s}{\to} g'$ and $s' \in A(g') - A(g)$, one has a real birth of s'. With this terminology in mind, equation (7.2.1) reads : the transition rate $x_g\lambda(s_0,g)$ out of g due to the death or pseudo-death of s_0 is equal to the transition rate into g with birth (real birth or pseudo-birth) of s_0, i.e. the transition rate into g with real birth of s_0 ($\sum_{g' \notin G_{s_0}} x_{g'} \sum_{s \in A(g')} p(g',s,g)\lambda(s,g')$) plus the transition rate into g with pseudo-birth of s_0 ($\sum_{g' \in G_{s_0}} x_{g'} p(g',s_0,g) \lambda(s_0,g)$). Of course, in the above interpretation, the transition rate is defined to be the average number of transitions per unit time.

7.3 Examples

The queue G/G/1/∞/LIFO preemptive

With the notation $x_{g_n} = x_n$, the global balance equations are

$$\begin{cases} x_0\lambda = x_1\mu, \\ x_n(\lambda+\mu) = x_{n+1}\mu + x_{n-1}\lambda \quad (n \geq 1). \end{cases} \tag{7.3.1}$$

If $\rho = \lambda/\mu < 1$ they admit a unique solution such that $\sum_{n=0}^{\infty} x_n = 1$, namely

$$x_n = (1-\rho)\,\rho^n \qquad (n \geq 0). \tag{7.3.2}$$

This solution verifies the insensitivity balance equations on all sites $n \geq 1$ since it can be checked that

$$x_n \lambda = x_{n+1} \mu \qquad (n \geq 1). \tag{7.3.3}$$

However, the insensitivity balance equation is not verified for site 0 because $x_0 \lambda \neq x_1\lambda$.

The G/G/1/∞/LIFO preemptive system is therefore insensitive on $S' = \mathbb{N}_+$ relatively to the intensities $(\lambda_s, s \in S) = (\lambda_n, n \in \mathbb{N})$ of the form $\lambda_0 = \lambda$, $\lambda_n = \mu$ $(n \geq 1)$. Since the only non insensitive site is the source, which generates customers, one also says : the queue M/G/1/∞/LIFO preemptive is insensitive.

The Erlang system G/G/k/0 :

The global balance equation is

$$x_g \, \Lambda(g) = \sum_{\substack{i=1 \\ i \notin g}}^{k} x_{g+\{i\}} \, \mu_i + \sum_{\substack{j=1 \\ j \in g}}^{k} x_{g-\{j\}} \, \lambda \, \pi(g-\{j\},j) + x_g \, \lambda \, \delta_{g,g_k} \qquad (7.3.4)$$

where $\delta_{g,g'}$ is the Kronecker symbol. Here $\Lambda(g) = \lambda + \sum_{i=1}^{k} \mu_i$.

Let now $i \in \{1,\dots,k\}$ and $i \in g$. The insensitivity balance equation relative to i and g is

$$x_g \, \mu_i = x_{g-\{i\}} \, \lambda \, \pi(g-\{i\},i). \qquad (7.3.5)$$

The insensitivity balance equation relatively to 0 and g_k is trivial : $x_g\lambda = x_g\lambda$. The insensitivity balance equation for 0 and $g \neq g_k$ is $x_g\lambda = 0$. If 0 were insensitive it should be verified for all g, thus 0 cannot be insensitive.

We will now see that, if the server assigned to a new customer is chosen randomly among the idle servers, i.e. if

$$\pi(g,i) = \frac{1}{k+1-|g|}, \qquad (7.3.6)$$

the insensitivity balance equations are satisfied on $S' = \{1,2,\dots k\}$. For this it suffices to check that the probability distribution

$$x_g = x_{g_0} \frac{\lambda^{|g|-1}}{\prod_{\substack{i=1 \\ i \in g}}^{k} \mu_i} \cdot \frac{(k-1)!}{(k+2-|g|)!}, \qquad (7.3.7)$$

where x_{g_0} is chosen in such a manner that $\sum_{g \in G} x_g = 1$, verifies both the global and the insensitivity balance equations.

7.4 Assumption on the input

The last assumption concerns the input (M_s, M_s, P_s) $(s \in S)$.

<u>A4 : For all $s \in S-S'$, $(M_s, \mathbf{M}_s, P)$ is a Poisson process.</u>

7.5 Two immediate consequences of the insensitivity balance equations

Restriction of birth for insensitive sites

(7.1.5) Let $g \in G$ and $s \in A(g) \cap S'$. Under assumptions A1, A2 and A3, either situation

(i) $g \overset{s}{\to} g'$, $s \in A(g')$, $|(A(g') - A(g)) \cap S'| \geq 1$

(ii) $g \overset{s}{\to} g'$, $s \notin A(g')$, $|(A(g') - A(g)) \cap S'| \geq 2$

implies that $p(g,s,g')\, c(s,g) = 0$.

In other words : At most one insensitive site can be born or pseudo-born at a given time.

Proof : Let s_1 and s_2 be two distinct insensitive sites and let g be a macrostate such that $\{s_1,s_2\} \subset A(g)$. Write equation (7.2.1) for $s_0 = s_1$, and $s_0 = s_2$, and substract the sum of the two equations so obtained from the global balance equation (7.1.5) to obtain

$$x_g \sum_{s\in\{s_1,s_2\}} \lambda(s,g) - \sum_{g'\in G_{s_1}\cap G_{s_2}} x_{g'} \sum_{s\in\{s_1,s_2\}} p(g',s,g)\,\lambda(s,g') =$$

$$- \sum_{g'\in \overline{G}_{s_1}\cap \overline{G}_{s_2}} x_{g'} \sum_{s\in A(g')} p(g',s,g)\,\lambda(s,g')$$

$$- \sum_{g'\in G_{s_1}\cap \overline{G}_{s_2}} x_{g'}\, p(g',s_1,g)\,\lambda(s_1,g')$$

$$- \sum_{g'\in \overline{G}_{s_1}\cap G_{s_2}} x_{g'}\, p(g',s_2,g)\,\lambda(s_2,g'). \qquad (7.5.2)$$

Let us make the summation of equations (7.5.2) when s ranges over $G_{s_1} \cap G_{s_2}$. The left-hand side of the sum is 0 (use (6.1.4)). The right-hand side terms of (7.5.2) are therefore zero, which was to be proved.

For future reference we shall need a variant of (7.5.2) : Substract from the global balance equation (7.1.5) the sum of equations (7.2.1) when so ranges over $S' \cap A(g)$, to obtain :

$$x_g \sum_{s\in A(g)\cap \overline{S}'} \lambda(s,g) = \sum_{\substack{g'\in \cap G_s \\ s\in A(g)\cap S'}} \quad \sum_{s'\in A(g')-S'} p(g',s',g)\,\lambda(s',g'), \qquad (7.5.3)$$

where $\overline{S}' = S-S'$.

Immediate activity of the insensitive sites

(7.5.4) Let $s_0 \in S'$ and $g \in G$, be such that $s_0 \in P(g)$ and consider any transition of the type : $g' \overset{s}{\rightarrow} g$. If $c(s_0,g) = 0$, then, either one of the two following situations

(i) $s = s_0$ and $s_0 \in P(g')$

(ii) $s_0 \notin AP(g')$ and $s \in A(g')$

implies that $p(g',s,g)\, c(s,g') = 0$.

In other words, if an insensitive site is born, it is immediately active.

Proof : Set $c(s_0,g) = 0$ in the insensitivity equation (7.2.1). The above result is just a consequence of the fact that all the right-hand terms of (7.2.1) are zero.□

8 THE INSENSITIVITY THEOREM

8.1 - The Palm version

If the probability P^0 defined in §6.3 verifies

$$P^0(X(0) = g,\ Y(0) \in B) = x_g^0\ \hat{P}_{A(g)}(B) \quad (g \in G,\ B \in \bigotimes_{s \in S} \mathbf{M}_s)$$

(where $\hat{P}_{A(g)}$ is given by (6.3.4)) and if $(x_g^0,\ g \in G)$ verifies the assumptions A1, A2, A3 and A4, the Markov chain $((X(T_n),\ \hat{Y}(T_n)),\ n \in \mathbb{N})$ is stationary with invariant probability measure

$$\pi^0(\{g\} \times \hat{B}) = x_g^0\ \hat{P}_{A(g)}(\hat{B}) \quad (g \in G,\ \hat{B} \in \bigotimes_{s \in S} (M_s \cap \hat{M}_s)). \tag{8.1.1}$$

In words, if there is insensitivity balance on S',the imbedded chain $(X(T_n),\ \hat{Y}(T_n))$ has π^0 for invariant measure whatever P_s, $s \in S'$. This will be proved in the Appendix.

From the stationarity of $\hat{Z}(T_n) = (X(T_n),\ \hat{Y}(T_n))$, it follows that the process

$$((X(T_{n-1}),\ \tilde{\sigma}_{n-1},\ X(T_n),\ \hat{Y}(T_n)),\ n \geq 1)$$

is also stationary and we shall see in the Appendix that

$$P^0(X(T_{n-1}) = g,\ \tilde{\sigma}_{n-1} = s,\ X(T_n) = g',\ \hat{Y}(T_n) \in \hat{B})$$

$$= x_g^0 \frac{\lambda(s,g)}{\Lambda(g)} p(g,s,g')\ P_{A(g)-\{s\}}(\hat{B}). \tag{8.1.2}$$

We shall also prove in the Appendix that

$$E^0[T_1] = \sum_{g \in G} \frac{x_g^0}{\Lambda(g)} \tag{8.1.3}$$

and therefore, by assumptions A1 and A2,

$$0 < E^0[T_1] < \infty. \tag{8.1.3'}$$

The important consequence of (8.1.1) is this. Taking $\hat{B}_s = \hat{M}_s$ $(s \in S)$ we find that

$$P^0(X(T_n) = g) = x_g^0 \quad (g \in G)\ (n \in \mathbb{N}). \tag{8.1.4}$$

The distribution of $X(T_n)$ $(n \in \mathbb{N})$ under P^0 therefore depends upon the law of $(M_s, \mathbf{M}_s, \mathbf{P}_s)$ $(s \in S)$ only through the intensities $(\lambda_s,\ s \in S)$ since it coincides with x_g^0. Recall however that for $s \in S'$, $(M_s, \mathbf{M}_s, \mathbf{P}_s)$ was assumed to be Poisson (assumption A4 of §7). This is why we shall be careful and say that the M/G/k/0 system is insensitive (not the G/G/k/0 system). Similarly M/G/∞ and M/G/1/∞/LIFO preemptive are insensitive.

Example

Let $U_s(T_n)$ be the residual time at site s and at time T_n. More precisely, if Φ_s is the initial point process attached to site s, then $U_s(T_n)$ is the first point strictly to the right of 0 of $T^s_{\theta_s(T_n)}$ (where $\theta_s(T_n) = \int_0^{T_n} c(s, X(u))du$). In zoological terms : at time T_n, the beast B_s sees the next point of the point process Φ_s at a distance $U_s(T_n)$.

Let $F_s^0(x)$ be the cumulative distribution function under P_s^0 of X_1^s (X_1^s is the position of the first point of Φ_s strictly to the right of 0, and therefore $X_1^s = U_s(0) = U_s(T_0)$).

Then formulae (8.1.1) and (6.3.4), and the results of §4.2 of Part I give

$$P^0(\{X(T_n)=g\} \quad \{U_{s_i}(T_n) \geq u_i,\ i \in \mathbb{N}^*\}) = x_g^0 \sum_{s \in A(g)} \frac{\lambda(s,g)}{\Lambda(g)} \prod_{i=1}^{\infty} H_i^{(s,g)}(u_i), \tag{8.1.5}$$

where

$$H_i^{(s,g)}(u_i) = \begin{cases} \lambda_{s_i} \int_{u_i}^{\infty} (1-F^0_{s_i}(x))dx & \text{if } s_i \in A(g) - \{s\} \\ 1 - F^0_{s_i}(u_i) & \text{if } s_i \notin A(g) - \{s\}, \end{cases} \tag{8.1.5}$$

provided the variable $(u_i,\ i \in \mathbb{N}^*)$ are zero but for a finite number of indices.

8.2 - From Palm to stationary

The situation is presently the following : we have under P^0 a stationary process $(\hat{Z}_n, n \in \mathbb{N})$ where $\hat{Z}_n = Z(T_n)$.

By a standard construction, we can extend this one-sided stationary sequence to a two-sided stationary sequence $(\hat{Z}_n, n \in \mathbb{Z})$. This involves in general a change of probability space, but we shall keep the same notations for the new space : $(\Omega, \mathbf{F}, P^0)$.

Each $\hat{Z}_n$ "contains" a random variable S_n which is the time to the next event, constructed from $\hat{Z}_n = (X_n, \hat{Y}_n)$ in the same manner as T_1 was constructed from $\hat{Z}_0 = (X_0, \hat{Y}_0) = (X(0), \hat{Y}(0))$ in § 6.4.

When we say that $\hat{Z}_n$ contains S_n, we mean that $S_n = f_n(\hat{Z}_n)$. In this particular situation f_n does not depend upon n. Thus $(S_n, \hat{Z}_n, n \in \mathbb{Z})$ is stationary under P^0. Now S_n generates a point process $(T_n, n \in \mathbb{Z})$ by

$$T_0 \equiv 0, \quad S_n = T_{n+1} - T_n \quad (n \in \mathbb{Z}) \tag{8.2.1}$$

It is a point process (i.e. $\lim_{n\to\infty} |T_n| = \infty$, P^0-a.s.) because we have made a regularity assumption (which is verified in virtually all queueing situations, as we have already seen in § 6.5.3).

All the above remarks will convince the reader that we can, at the expense of transporting $(\Omega, \mathbf{F}, P^0)$ into some canonical space of marked point processes, assume that the setting is the following :

There exists a probability space $(\Omega, \mathbf{F})$ endowed with a semi-group of measurable automorphisms $(\theta_t, t \in \mathbb{R})$ on which is defined :

- A point process N such that $N \circ \theta_t = T_t N$ $(t \in \mathbb{R})$, of basic points $(T_n, n \in \mathbb{Z})$.
- Sequences of marks $\hat{Z} = (\hat{Z}_n, n \in \mathbb{Z})$, where $\hat{Z}_n = \hat{Z}_0 \circ \theta_{T_n}$, such that $\hat{Z}_{n+1}$ is obtained from $\hat{Z}$ as $\hat{Z}(T_1)$ is obtained from $\hat{Z}(T_0)$ in §6.3.
- A probability P_N^0 such that $P_N^0(T_0 = 0) = 1$, $((X_n, \hat{Y}_n), n \in \mathbb{Z})$ is stationary under P_N^0, and $P_N^0((X_n, \hat{Y}_n) \in \{g\} \times \hat{B}) = x_g \hat{P}^0_{A(g)}(\hat{B})$ for all $g \in G$, $\hat{B} \in \bigotimes_{s \in S} (\mathbf{M}_s \cap \hat{M}_s)$.

Moreover P_N^0 is θ_{T_1}-stationary.

In this new setting, and in view of (8.1.3') which now reads $E_0^N[T_1] < \infty$, it is clear that P_N^0 is the Palm probability with respect to N of a θ_t-stationary probability measure P.

8.3 – The stationary version and Matthes Product Form

We recall the definition of $\hat{Z}(t) = (X(t), \hat{Y}(t))$ for all $t \in \mathbb{R}$:

$$\begin{cases} X(t) = X(T_n) = X_n & \text{if } t \in [T_n, T_{n+1}), \\ \hat{Y}(t) = \left(\overbrace{T^s_{\int_{T_n}^{t} c(s,X(u))du} \qquad \hat{Y}^s_n}, \; s \in S \right) & \text{if } t \in [T_n, T_{n+1}), \end{cases} \tag{8.3.1}$$

where $(\hat{Y}^s_n, s \in S) = \hat{Y}_n$. The latter awesome formalism says that $\hat{Y}(t)$ is what remains of $\hat{Y}_n = Y(T_n)$ at time t knowing that each beast B_s has been eating its part $\hat{Y}^s_n$ of Y_s at speed $c(s,X(u))$ for all $u \in [T_n,t]$ where $t \in [T_n, T_{n+1})$.

One much check that definition (8.3.1) is consistent, in particular that $\hat{Y}(T_n) = \hat{Y}_n$ for all $n \in \mathbb{Z}$. This is true because

$$\hat{Y}(t) = \left(\overbrace{T^s_{\int_{T_n}^{T_{n+1}} c(s,X(u))du} \qquad \hat{Y}^s_n}, \; s \in S \right) \quad (n \in \mathbb{Z})$$

is true P^0_N-a.s., and therefore P-a.s. (cf. §8.1 of Part 1, and §2.1)

The last verification shows that this part of the evolution process can be described in terms of the "eating mechanism" with respect to probability P as well. Therefore as far as this part of the Bedienung mechanism is concerned, the passage from P^0_N to P does not change anything. There is another aspect of the mechanism which we must now check : are the transition $g \overset{s}{\to} g'$ governed under P as they were under P^0_N ? In view of the theoretical nature of the passage from P^0_N to P, this is not immediately obvious.

We must show

$$P(X(T_{n+1}) = g' \mid X(T_n) = g, \tilde{\sigma}_n = s) = p(g,s,g'), \tag{8.3.2}$$

where $\tilde{\sigma}_n$ is the site on which is created event T_{n+1}. The inversion formula entails

$$P(X(T_{n+1}) = g', X(T_n) = g, \tilde{\sigma}_n = s) =$$

$$= \frac{1}{E^0[T_1]} \int_0^\infty P^0_N(u \le T_1, X(T_{n+1} \circ \theta_u) = g', X(T_n \circ \theta_u) = g, \tilde{\sigma}_n \circ \theta_u = s) du$$

$$= \frac{1}{E^0[T_1]} \int_0^\infty P^0_N(u \le T_1, X(T_{n+1}) = g', X(T_n) = g, \tilde{\sigma}_n = s) du.$$

In addition,

$P_N^0\,(u \le T_1,\ X(T_{n+1}) = g',\ X(T_n) = g,\ \tilde{\sigma}_n = s) =$

$P_N^0\,(u \le T_1,\ X(T_n) = g,\ \tilde{\sigma}_n = s)\ P_N^0\,(X(T_{n+1}) = g' \mid u \le T_1,\ X(T_n) = g,\ \tilde{\sigma}_n = s) =$

$P_N^0\,(u \le T_1,\ X(T_n) = g,\ \tilde{\sigma}_n = s)\ p(g,s,g')$.

Therefore

$$P(X(T_{n+1}) = g',\ X(T_n) = g,\ \tilde{\sigma}_n = s) = f_n(s,g)\ p(g,s,g'), \qquad (8.3.3)$$

for some function f. The latter equality implies (8.3.2). ∎

It is clear that

$$Z(0)\circ\theta_t = Z(t), \qquad (t \in \mathbb{R}). \qquad (8.3.4)$$

Therefore $(\hat{Z}(t),\ t \in \mathbb{R})$ is stationary under P.

We will establish in the Appendix the following formula (the so called Matthes product form)

$$P(X(0) = g,\ \hat{Y}(0) \in \hat{B}) = x_g\ P_{A(g)}(B), \qquad (g \in G,\ \hat{B} \in \mathbf{M}_s \cap \hat{M}_s). \qquad (8.3.5)$$

9 - INSENSITIVITY BALANCE EQUATIONS ARE NECESSARY FOR INSENSITIVITY

9.1 The converse theorem

Let $\Sigma = \{G,S,c,p\}$ be a mechanism, S' be a subset of S and $\Lambda = \{\lambda_s,\ s \in S\}$ a fixed vector of intensities. All the inputs $\{(M_s,\mathbf{M}_s,P_s),\ s \in S\}$ to be considered in this section will be supposed to satisfy assumption (A.4) and to be such that the intensity of $(M_s,\mathbf{M}_s,P_s)$ is λ_s, $s \in S$ above.

It will be assumed that the infinitesimal generator given by (7.1.6) is of a regular and irreducible continuous time Markov chain with state space G. Both this continuous time Markov chain and its imbedded chain will also be assumed to be ergodic (assumptions A.1 and A.2) and we shall denote as $\{x_g,\ g \in G\}$ the unique solution of the global balance equations (7.1.5).

In view of the results of section 8, it makes sense to adopt the following definition :

S' is a set of insensitive sites if for all input $(M_s,\mathbf{M}_s,P_s)$, $s \in S$, satisfying (A.4), there is an initial generalized state distribution such that $(Z(t),\ n \in \mathbb{Z})$ is stationary on $(\Omega,F^0,P_N^0,\theta)$, $0 < E_{P_N^0}[T_1] < \infty$ and such that the corresponding time stationary version satisfies the condition

$$P[X(t) = g] = x_g, \quad g \in G. \qquad (9.1.1)$$

The aim of this section is to prove the following converse to the insentivity theorem of section 8 :

(9.1.2) <u>If S' is a set of insensitive sites for $(\sum,\Lambda)$ and if the sites of S' satisfy the property of immediate activity (§ 7), then all the sites of S' satisfy the insensitivity balance equation (7.2.1).</u>

9.2 The method of stages

To prove (9.1.2) for $s_0 \in S'$, consider the following inputs : $(M_{s_0}, \mathbf{M}_{s_0}, P_{s_0})$ is a renewal process characterized by the following distribution function

$$F(t) = \Pi\, E_{\lambda_1}(t) + (1-\Pi)\, E_{\lambda_2} * E_{\lambda_1}(t),\ t \geq 0, \tag{9.2.1}$$

where E_λ denotes the exponential distribution function of parameter λ and * the convolution. The constants Π, λ_1, λ_2 are assumed to be positive, with $0 < \Pi < 1$ and to satisfy the following intensity conservation equation

$$\int_0^\infty t\,dF(t) = \lambda_{s_0}^{-1}\ . \tag{9.2.2}$$

Let σ_1 and σ_2 be two new sites. With σ_i is associated an independent Poisson process $(M_{\sigma_i}, \mathbf{M}_{\sigma_i}, P_{\sigma_i})$ with intensity λ_i, $i = 1,2$. The time it takes for B_{s_0} to gnaw one interarrival time of m_{s_0} distributed according to F(t) can be represented by a random variable equal with probability Π to the time it takes for B_{σ_1} to gnaw one exponential random variable of parameter λ_1 and with probability $(1-\Pi)$ to the total time it takes for B_{σ_2} to gnaw one exponential variable of parameter λ_2 and for B_{σ_1} one of parameter λ_1. This representation in terms of "stages" leads to the definition of a new equivalent mechanism $\hat{\sum} = \{\hat{G}, \hat{S}, \hat{c}, \hat{p}\}$ with Poisson inputs on all its sites : take

$$\hat{S} = (S - \{s_0\}) \cup \{\sigma_1\} \cup \{\sigma_2\} \tag{9.2.3}$$

and

$$\hat{G} = \{g \notin G_{s_0}\} \cup \{\sigma_i(g),\ g \in G_{s_0},\ i = 1,2\}, \tag{9.2.4}$$

where $\sigma_1(g)$ and $\sigma_2(g)$ are two new macrostates such that $P(\sigma_i(g)) = (P(g) - \{s_0\}) \cup \{\sigma_i\}$, $i = 1,2$, $g \in G_{s_0}$). The speeds $\hat{c}(.,.)$ are defined by

$$\begin{cases} \hat{c}(.,g) = c(.,g) & \text{for } g \notin G_{s_0}, \\ \hat{c}(s,\sigma_i(g)) = c(s,g) & \text{for } g \in G_{s_0},\ s \neq \sigma_i, \\ \hat{c}(\sigma_i,\sigma_i(g)) = c(s_0,g) & \text{for } g \in G_{s_0}, \end{cases} \tag{9.2.5}$$

whence the transition kernel $\hat{p}$ is defined by

$$\begin{cases} \hat{p}(g,.,g') = p(g,.,g') & ,g,g' \notin G_{s_0}, \\ \left.\begin{array}{l} \hat{p}(g,.,\sigma_1(g')) = p(g,.,g').\Pi \\ \hat{p}(g),.,\sigma_2(g')) = p(g,.,g')(1-\Pi) \end{array}\right\} & ,g \in G_{s_0},\ g' \in G_{s_0}, \\ \hat{p}(\sigma_i(g),s,\sigma_i(g')) = p(g,s,g') & ,g,g' \in G_{s_0},\ s \notin \{\sigma_1,\sigma_2\}, \\ \hat{p}(\sigma_2(g),\sigma_2,\sigma_1(g)) = 1 & ,g \in G_{s_0}, \\ \hat{p}(\sigma_1(g),\sigma_1,g') = p(g,s_0,g') & ,g \in G_{s_0},\ g' \notin G_{s_0} \\ \left.\begin{array}{l} \hat{p}(\sigma_1(g),\sigma_1,\sigma_1(g') = p(g,s_0,g')\Pi \\ \hat{p}(\sigma_1(g),\sigma_1,\sigma_2(g')) = p(g,s_0,g')(1-\Pi) \end{array}\right\} & ,g,g' \in G_{s_0}. \end{cases} \tag{9.2.6}$$

Let $\hat{\Lambda} = \{\hat{\lambda}_s, s \in \hat{S}\}$ be the set of intensities

$$\begin{cases} \hat{\lambda}_s = \lambda_s \text{ for } s \in S - \{s_0\}, \\ \hat{\lambda}_{\sigma_i} = \lambda_i. \end{cases} \tag{9.2.7}$$

The reason for introducing this specific law and this new mechanism is that it allows to separate between the transitions out of $g \in G_{s_0}$ due to an event of s_0 (which can only take place from $\sigma_1(g)$) and the other transitions. This becomes apparent in the global balance equations for $(\hat{\Sigma},\hat{\Lambda})$ which read as follows for $g \in G_{s_0}$ (equations (9.2.8) - (9.2.9)) :

$$\hat{x}_{\sigma_1(g)}(\Lambda(g) - \lambda(s_0,g) + \lambda_1(s_0,g)) = \sum_{g' \in G_{s_0}} \hat{x}_{\sigma_1(g')} \sum_{s \neq s_0} p(g',s,g)\,\lambda(s,g')$$

$$+ \hat{x}_{\sigma_2(g)}\,\lambda_2(s_0,g) + \Pi \sum_{g' \notin G_{s_0}} \hat{x}_{g'} \sum_{s \in A(g')} p(g',s,g)\,\lambda(s,g')$$

$$+ \Pi \sum_{g' \in G_{s_0}} \hat{x}_{\sigma_1(g')}\, p(g',s_0,g)\,\lambda_1(s_0,g'), \tag{9.2.8}$$

where $\lambda_i(s_0,g) = \lambda_i\ c(s_0,g)$ and

$$\hat{x}_{\sigma_2(g)}\,(\Lambda(g) - \lambda(s_0,g) + \lambda_2(s_0,g)) = \sum_{g' \in G_{s_0}} \hat{x}_{\sigma_2(g')} \sum_{s \neq s_0} p(g',s,g)\ \lambda(s,g')$$

$$+ (1-\Pi) \sum_{g' \notin G_{s_0}} \hat{x}_{g'} \sum_{s \in A(g')} p(g',s,g)\ \lambda(s,g')$$

$$+ (1-\Pi) \sum_{g' \in G_{s_0}} \hat{x}_{\sigma_1(g')}\ p(g',s_0,g)\ \lambda_1(s_0,g'). \qquad (9.2.9)$$

For $g \notin G_{s_0}$, global balance reads

$$\hat{x}_g\ \Lambda(g) = \sum_{g' \notin G_{s_0}} \hat{x}_{g'} \sum_{s \in A(g')} p(g',s,g)\ \lambda(s,g')$$

$$+ \sum_{g' \in G_{s_0}} \hat{x}_{\sigma_1(g')}\ p(g',s_0,g)\ \lambda_1(s_0,g'). \qquad (9.2.10)$$

The first consequence of the insensilivity assumption is that

(9.2.11) <u>The global balance equations (9.2.8)-(9.2.10) have a unique solution $\{\hat{x}_g, g \in \hat{G}\}$.</u>

Let $(\hat{X}(t),\ t \geq 0)$ be the continuous time Markov process with state space $\hat{G}$ associated with the global balance equations above. $\hat{X}(t)$ is irreducible. This property stems from the irreducibility assumption on the Markov process with state space G (equation (7.1.6)).

Also, for virtually all systems, $\hat{X}(t)$ is regular (see (6.5.10)) and therefore this assumption will be made.

We shall prove that $\hat{X}(t)$ is ergodic. Indeed if it were not, for some g and g' in $G-G_{s_0}$,

$$\lim_{t\to\infty} P[\hat{X}(t) = g \mid \hat{X}(0) = g'] = 0.$$

By construction, we have

$$P[\hat{X}(t) = g \mid \hat{X}(0) = g'] = P[X(t) = g \mid X(0) = g'],$$

where $X(t)$ is the first component of the generalized state process $Z(t)$ (9.1.1).

Hence, we would have

$$\lim_{t\to\infty} P[X(t) = g \mid X(0) = g'] = 0,$$

which contradicts the assumption that $Z(t)$ can be made stationary.

Insensitivity also implies the following relations :

$$\hat{x}_g = x_g, \quad \text{for } g \notin G_{s_0} \tag{9.2.12}$$

and

$$\hat{x}_{\sigma_1(g)} + \hat{x}_{\sigma_2(g)} = x_g, \text{ for } g \in G_{s_0}. \tag{9.2.13}$$

9.3 Proof of the converse theorem

The first step consists in proving that the relation

$$\hat{x}_{\sigma_1(g)} \lambda_1 = x_g \lambda_{s_0} \tag{9.3.1}$$

holds true for all $g \in G_{s_0}$ such that $c(s_0,g) \neq 0$: the intensity of the transitions out of g due to an event on s_0 is insensitive. Notice that (9.3.1), (9.2.2) and (9.2.13) entail that if $c(s_0,g) \neq 0$,

$$\hat{x}_{\sigma_2(g)} \lambda_2 = x_g(1-\Pi) \lambda_{s_0}, \ g \in G_{s_0}. \tag{9.3.2}$$

The global balance equation for (Σ,Λ) reads

$$x_g \Lambda(g) = \sum_{g' \in G} x_{g'} \sum_{s \in A(g')} p(g',s,g) \lambda(s,g'). \tag{9.3.3}$$

Using (9.2.12) and (9.2.13), we get the following relation from the term by term operation (9.2.8)+(9.2.9)-(9.3.3) :

$$a(g) = \sum_{g' \in G_{s_0}} a(g') p(g',s_0,g), \ g \in G_{s_0}, \tag{9.3.4}$$

where

$$a(g) = c(s_0,g) (\lambda_1 \hat{x}_{\sigma_1(g)} - \lambda_{s_0} x_g), \ g \in G_{s_0}. \tag{9.3.5}$$

In view of the ergodicity assumption on the imbedded Markov chain of (Σ,Λ), we have

$$\sum_{g \in G} x_g \Lambda(g) < \infty, \tag{9.3.6}$$

so that

$$\sum_{g \in G_{s_0}} |a(g)| < \infty. \tag{9.3.7}$$

In the case where the kernel $p(g,s_0,g') : G_{s_0} \to G_{s_0}$ is irreducible and transient (for instance sub-Markov), we get immediately that $a(g) = 0, \forall g \in G_{s_0}$, from (9.3.4) and (9.3.7). The proof for the general case is postponed to Appendix 5.

If the condition $c(s_0,g) > 0$ is satisfied for all $g \in G_{s_0}$, the relations (9.3.1) and (9.3.2) giving $\hat{x}_{\sigma_1(g)}$ in terms of x_g are valid for all $g \in G_{s_0}$. Using these relations in the global balance equation (9.2.9), we obtain (after division of both sides by $(1-\Pi)$)

$$x_g \frac{\lambda_{s_0}}{\lambda_2}(\Lambda(g) - \lambda(s_0,g) + \lambda_2(s_0,g)) = \sum_{g' \in G_{s_0}} x_{g'} \frac{\lambda_{s_0}}{\lambda_2} \sum_{s \neq s_0} p(g',s,g)\,\lambda(s,g')$$

$$+ \sum_{g' \in G_{s_0}} x_{g'}\, p(g',s_0,g)\,\lambda(s_0,g') + \sum_{g' \notin G_{s_0}} x_{g'} \sum_{s \in A(g')} p(g',s,g)\,\lambda(s,g'). \tag{9.3.8}$$

This equation and the global balance equation of (Σ,Λ) for g imply the relation

$$x_g(\Lambda(s_0,g)-\lambda(s_0,g)) - \sum_{g' \in G_{s_0}} x_{g'} \sum_{s \neq s_0} p(g',s,g)\,\lambda(s,g') = 0, \tag{9.3.9}$$

provided $\lambda_{s_0} \neq \lambda_2$, which will be assumed from now on. Inserting (9.3.4) in the global balance equation of (Σ,Λ) for g, we finally obtain

$$x_g\,\lambda(s_0,g) = \sum_{g \notin G_{s_0}} x_{g'} \sum_{s \in A(g')} p(g',s,g)\,\lambda(s,g')$$

$$+ \sum_{g' \in G_{s_0}} x_{g'}\, p(g',s_0,g)\,\lambda(s_0,g'),\ g \in G_{s_0}. \tag{9.3.10}$$

This is <u>the insensitivity balance equation for (g,s_0)</u>. The proof in the general case (i.e. with possibly vanishing speeds) requires the <u>immediate activity assumption</u> and will be given in Appendix 6.

9.4 Example M/G/1/∞/FIFO is not insensitive

Consider the following mechanism representing a G/G/1/∞/FIFO queue. Let $S=\{s_0,s_1\}$ and $G = \{g_0,g_1,\ldots,g_k,\ldots\}$. Take $A(g_k) = \{s_0,s_1\}$ for $k \geq 1$ and $A(g_0)=\{s_0\}$. The transition kernel is defined by

$$\begin{cases} p(g_k, s_0, g_{k+1}) = 1, & \text{for } k \geq 0, \\ p(g_k, s_1, g_{k-1}) = 1, & \text{for } k \geq 1. \end{cases} \qquad (9.4.1)$$

and the speeds by

$$\begin{cases} c(s_0,g_k) = 1 & \text{for } k \geq 0, \\ c(s_1,g_k) = 1 & \text{for } k \geq 1. \end{cases} \qquad (9.4.2)$$

Let λ and μ be the intensities associated with s_0 and s_1 respectively.

Let $\rho = \lambda/\mu$. If $\rho < 1$, the global balance equations are known to have for unique solution

$$x_{g_k} = \rho^k(1-\rho), \quad k \geq 0. \qquad (9.4.3)$$

Notice that this mechanism has positive speeds. Using the converse theorem of section 9.1, we conclude that s_1 is not an insensitive site for this mechanism. Indeed the insensity balance equation for (g_k,s_1) is not satisfied for $k\geq 1$. Hence this representation of the M/G/1/∞ FIFO queue is not insensitive with respect to the service law.

A stronger result can actually be derived from the converse theorem :

(9.4.4) <u>There exists no mechanism representing M/G/1/∞/FIFO and insensitive with respect to the service law.</u>

Any other mechanism Σ' representing this system has to be such that some aggregation Γ'_k of its state space G' coincides with the event "There are k customers in the queue", k=0,1,... Since the probability of Γ'_k in Σ' has to coincide with the probability of g_k in Σ, the sensitivity of Σ with respect to changes on P_{s_1} entails the sensitivity of Σ' with respect to changes on the service law, whatever the representation of services in Σ'.

10 - POISSON STREAMS

10.1 Privileged transitions

The situation is that of sections 6, 7, 8 and 9. The main features of interest in the present section are the existence of a stationary point process (N,θ_t,P) where $N = \sum_{n\in\mathbb{Z}} \delta_{T_n}$ and of a sequence of marks $(U_n,\ n \in \mathbb{Z})$ where $U_n=(X_{n-1},\ \tilde{\sigma}_{n-1},\ X_n,\hat{Y}_n) \in G\times S\times G\times(\prod_{s\in S} \hat{M}_s)$. Recall that $X_n = X(T_n)$ is the state at time T_n, and $\tilde{\sigma}_{n-1}$ is the site of $A(X_{n-1})$ which triggers the transition $X_{n-1} \to X_n$, i.e. $X_{n-1} \xrightarrow{\tilde{\sigma}_{n-1}} X_n$. If P_N^0 is the Palm probability associated with (N,θ_t,P), we have (Eq. (8.1.2)) :

$$P_N^0(X_{n-1}=g,\ \tilde{\sigma}_{n-1}=s,\ X_n=g',\ \hat{Y}_n \in B) = \frac{\lambda(s,g)}{\Lambda(g)} p(g,s,g')\ x_g^0\ P_{A(g)-\{s\}}(\hat{B}) \qquad (10.1.1)$$

Let

$$H \subset G\times S\times G \qquad (10.1.2)$$

be a subset of <u>privileged transitions</u> $g \overset{s}{\to} g'$.

Let now N_H be the point process associated with H, i.e.

$$N_H(.) = \sum_{n\in\mathbb{Z}} 1_{\{(X_{n-1},\tilde{\sigma}_{n-1},X_n)\in H\}}\ \delta_{T_n}(.) \qquad (10.1.3)$$

The Palm probability $P_{N_H}^0$ associated with (N_H,θ_t,P) is (see Part 1, §5.2)

$$P_{N_H}^0(A) = \frac{P_N^0(A\cap\{(X_0,\tilde{\sigma}_0,X_1)\in H\})}{P_N^0((X_0,\tilde{\sigma}_0,X_1)\in H)}\ . \qquad (10.1.4)$$

From (10.1.1)

$$P_N^0((X_0,\tilde{\sigma}_0,X_1) \in H) = \sum_{(g,s,g')\in H} \frac{\lambda(s,g)}{\Lambda(g)}\ x_g^0\ p(g,s,g'). \qquad (10.1.5)$$

Calling Λ_H the right-hand side of (10.1.5) and applying (10.1.4), we have

$$P_N^0(X_{-1}=g,\ \tilde{\sigma}_{-1}=s,\ X_0=g',\ \hat{Y}_0 \in \hat{B}) =$$

$$1_H(g,s,g')\ \frac{\lambda(s,g)}{\Lambda(g)} p(g,s,g')\ P_{A(g)-\{s\}}(\hat{B})/\Lambda_H \qquad (10.1.6)$$

and therefore

$$P_{N_H}^0(X_0=g',\hat{Y}_0\in\hat{B}) = \sum_{(g,s)\in G\times S} 1_H(g,s,g')\ \frac{\lambda(s,g)}{\Lambda(g)} p(g,s,g')\ P_{A(g)-\{s\}}(\hat{B})/\Lambda_H. \quad (10.1.7)$$

Let us recall that

$$P(X_0=g', \hat{Y}_0 \in \hat{B}) = x_{g'} \hat{P}_{A(g')}(\hat{B}). \tag{10.1.8}$$

10.2 Sufficient conditions for Poissonian streams

Suppose that for all $g' \in G$, $\hat{B} \in \left(\bigotimes_{s \in S} (M_s \cap \hat{M}_s)\right)$, the right-hand sides of (10.1.7) and (10.1.8) are equal. It is then clear that the restrictions of P and $P^0_{N_H}$ to $\sigma(T^H_n ; n \geq 1)$ are equal, where $(T^H_{N}, n \in \mathbb{Z})$ is the sequence of points of N_H. From the characterization of Poisson processes of §7, Part 1, it follows that N_H is a Poisson process.

The following condition is sufficient for this to happen :

$$(g,s,g') \in H \Rightarrow A(g') \cap S' = (A(g) - \{s\} \cap S'), \tag{10.2.1a}$$

$$x_{g'} = \frac{\sum_{(s,g) \in S \times G} 1_H(g,s,g')\, x_g\, \lambda(s,g)\, p(g,s,g')}{\sum_{(g,s,g') \in H} x_g\, \lambda(s,g)\, p(g,s,g')} . \tag{10.2.1b}$$

Indeed (10.2.1a) implies that for all $\hat{B} \in \bigotimes_{s \in S} (M_s \cap \hat{M}_s)$

$$P_{A(g)-\{s\}}(\hat{B}) = \hat{P}_{A(g')}(\hat{B}) , \text{ if } (g,s,g') \in H \tag{10.2.2}$$

and (10.2.1b) concludes the proof. ◻

APPENDIX

Appendix 1. Change of scale

With each macrostate $g \in G$, we associate a measurable flow $(T_t^g,\ t \in \mathbb{R})$ on $(\prod_{s\in S} M_s,\ \bigotimes_{s\in S} \mathcal{M}_s)$ by

$$T_t^g((m_s,\ s \in S)) = (T^s_{tc(s,g)}\, m_s,\ s \in S). \qquad (A.1.1)$$

Note that if $s \notin A(g)$, $c(s,g) = 0$, so that $T^s_{tc(s,g)}\, m_s = m_s$.

Define the following r.v. on $(\prod_{s\in S},\ \bigotimes_{s\in S} \mathcal{M}_s)$

$$\tau = \inf \left\{\frac{x_1(m_s)}{c(s,g)};\ s \in A(g)\right\}, \qquad (A.1.2)$$

where $x_1(m_s)$ is the first point of m_s strictly to the right of 0.

Lemma : $\hat{P}_{A(g)}$ is T^g_τ-invariant.

Proof : Define the mapping $k_{s,g} : M_s \to M_s$ by

$$k_{s,g}(m_s)(.) = \sum_{i\in\mathbb{Z}} \delta_{\frac{x_i(m_s)}{c(s,g)}}(.). \qquad (A.1.3)$$

This mapping performs a change of scale, so that if we denote $\Pi_{s,g} = P_s o k^{-1}_{s,g}$, $(M_s, \mathcal{M}_s, \Pi_{s,g})$ is a stationary point process with the intensity $\lambda(s,g) = \lambda_s c(s,g)$. If $s \in A(g)$, $\lambda(s,g) > 0$, and the Palm probability $\Pi^0_{s,g}$ associated with $\Phi_s, \Pi_{s,g}$ is $P^0_s \ o\ k^{-1}_{s,g}$.

Define the probability $\tilde{Q}$ on $(\prod_{s\in A(g)} M_s,\ \bigotimes_{s\in A(g)} \mathcal{M}_s)$ by

$$\tilde{Q} = \bigotimes_{s\in A(g)} \Pi_{s,g}. \qquad (A.1.4)$$

The Palm probability $\tilde{Q}^\circ$ of $\tilde{Q}$ with respect to

$$N = \sum_{s\in A(g)} m_s \qquad (A.1.5)$$

is given by

$$\tilde{Q}^{\circ} = \sum_{\ell=1}^{|A(g)|} \frac{\lambda(u_\ell, g)}{\Lambda(g)} \left(\bigotimes_{k=1}^{\ell-1} \Pi_{u_k,g} \right) \otimes \Pi^{\circ}_{u_\ell,g} \times \left(\bigotimes_{k=\ell+1}^{|A(g)|} \Pi_{u_k,g} \right), \tag{A.1.6}$$

where $\{u_1,\ldots,u_{|A(g)|}\}$ is an enumeration of A(g). Clearly $\tilde{Q}^{\circ}$ is $T_{\tau'}$-invariant, where τ' is the r.v. defined on $(\prod_{s\in A(g)} M_s, \bigotimes_{s\in A(g)} M_s)$ by

$$\tau' = \inf\{x_1(m_s),\ s \in A(g)\} \tag{A.1.7}$$

and $T_t((m_s,\ s \in A(g)) = (T_t^s\, m_s,\ s \in A(g))$.

Denote by k_g the mapping from $\prod_{s\in A(g)} M_s$ into itself defined by $k_g((m_s,\ s \in A(g)))$ $= (k_{s,g}(m_s),\ s \in A(g))$. If $\hat{\tau}$ is the restriction of τ to $(\prod_{s\in A(g)} M_s, \bigotimes_{s\in A(g)} M_s)$, and if $\hat{T}^g_t$ is the restriction of T^g_t to this space, then $\hat{T}^g_\tau$ is the image of $T_{\tau'}$ by k_g. Therefore the probability

$$Q^0 = \sum_{\ell=1}^{|A(g)|} \frac{\lambda(u_\ell, g)}{\Lambda(g)} \left(\bigotimes_{k=1} P_{u_k} \right) \otimes P^0_{u_\ell} \otimes \left(\bigotimes_{k=\ell+1}^{|A(g)|} P_{u_k,g} \right) \tag{A.1.8}$$

is $\hat{T}^g_\tau$-invariant.

The lemma is now established since Q° is the restriction of $\hat{P}_{A(g)}$ to $(\prod_{s\in A(g)} M_s, \bigotimes_{s\in A(g)} M_s)$, and since $T^g_t m_s = m_s$ when $s \notin A(g)$. □

Remark 1 : The point process N defined by (A.1.5) admits the intensity $\Lambda(g)$ under Q, and the inversion formula giving $\tilde{Q}$ in terms of $\tilde{Q}^{\circ}$ is

$$\tilde{Q}(A) = \Lambda(g)\ E_{\tilde{Q}^{\circ}} \left[\int_0^{\tau'} 1((T_u m_s,\ s \in A(g)) \in A)\, du\right], \quad (A \in \bigotimes_{s\in A(g)} M_s). \tag{A.1.9}$$

Therefore, since $\tilde{Q} = Q o k_g^{-1}$ and $\tilde{Q}^{\circ} = Q^{\circ}\ o\ k_g^{-1}$,

$$Q(A) = \Lambda(g)\ E_{Q^{\circ}} \left[\int_0^{\hat{\tau}} 1((\hat{T}^g_u m_s,\ s \in A(g)) \in A)\, du\right), \quad (A \in \bigotimes_{s\in A(g)} M_s). \tag{A.1.10}$$

Remark 2 : We have

$$E_{\hat{P}_{A(g)}}[\tau] = E_{\tilde{Q}_0}[\tau']. \tag{A.1.11}$$

Therefore in view of the construction of $(Z(t),\ t \in \mathbb{R})$ in §6, and of the definition of T_1

$$E_{P^0}[T_1] = \sum_{g\in G} x_0^g\, E_{\hat{P}_{A(g)}}[\tau] = \sum_{g\in G} \frac{x_g^0}{\Lambda(g)} < \infty, \tag{A.1.13}$$

where the last inequality follows from assumption A.2 of §7. Using (7.1.4) we obtain

$$E_{P^0}[T_1] = K \sum_{g\in G} x_g = K \tag{A.1.14}$$

Appendix 2. Proof of insensitivity (Proof of (8.1.1))

We adopt the notation $A(g) = \{u_1,\dots u_{|A(g)|}\}$. Let $g \in G$ and $B \in \bigotimes_{s\in S} \hat{m}_s$ be fixed. C_k is the event

$$C_k^g = (T_\tau^g)^{-1}\,(B \cap \{x_0(m_{u_k}) = 0\},$$

where $x_0(m_s)$ is the first point of m_s less or equal to zero. Considering all the possible values for $\tilde{\sigma}_0$, we get

$$P^0[X(0) = g,\ X(T_1) = g',\ Y(T_1) \in B]$$

$$= \sum_{k=1}^{|A(g)|} P^0[X(0) = g,\ X(T_1) = g',\ Y(T_1) \in B \cap \{x_0(m_{u_k}) = 0\}]$$

$$= \sum_{k=1}^{|A(g)|} P^0[X(0) = g,\ Y(0) \in C_k^g,\ X(T_1) = g'].$$

Using the conditional independence of $X(T_1)$ and $Y(0)$ knowing $(X(0),\tilde{\sigma}_0)$ (see §6.4), we can rewrite the last expression as

$$\sum_{k=1}^{|A(g)|} P^0[X(0) = g,\ Y(0) \in C_k^g].p(g,u_k,g') = \sum_{k=1}^{|A(g)|} x_g^0\, \hat{P}_{A(g)}(C_k^g).p(g,u_k,g'),$$

where the R.H.S. is obtained from the very definition of P^0 (see (6.3.6)).

But in view of the lemma of Appendix 1,

$$\hat{P}_{A(g)}(C_k^g) = \hat{P}_{A(g)}(T_\tau^g(C_k^g)) = \hat{P}_{A(g)}(B\cap\{x_0(m_{u_k}) = 0\}).$$

Since

$$\hat{P}_{A(g)}(B\cap\{x_0(m_{u_k})=0\} = P_{A(g)-\{u_k\}}(B)\ \frac{\lambda(u_k,g)}{\Lambda(g)},$$

we finally get

$$P(X(0)=g,\ X(T_1)=g',\ Y(T_1)\in B) = \sum_{k=1}^{|A(g)|} x_g^0\ p(g,u_k,g')P_{A(g)-\{u_k\}}(B).\frac{\lambda(u_k,g)}{\Lambda(g)}. \tag{A.2.1}$$

For any $u \in A(g)$ such that $g \overset{u}{\to} g'$ is allowed, we have

$$P_{A(g)-\{u\}}(B) = P_{A(g')-\{v\}}(B)\ \text{if}\ (A(g')-A(g))\cap S' = \{v\}. \tag{A.2.2}$$

To show this, use the birth restriction property of §7.5, which implies that if u is immediately reactivated after the transition $g \overset{u}{\to} g'$, then necessarily $u \in \bar{S}'$ since $v \in S'$; also use the property that for Poisson processes stationary probability and Palm probability coincide on all events implying only the restriction to $\mathbb{R}^+ - \{0\}$ of the point process (see §7 of Part 1). We have similarly

$$P_{A(g)-\{u\}}(B) = P_{A(g')}(B)\ \text{if}\ A(g')\cap S' \subset A(g)\ \text{and}\ u \notin A(g') \tag{A.2.3}$$

and

$$P_{A(g)-\{u\}}(B) = P_{A(g')-\{u\}}(B)\ \text{if}\ A(g')\cap S' \subset A(g)\ \text{and}\ u \in A(g'). \tag{A.2.4}$$

Let now $u \in A(g')\cap S'$. Write the local balance equation for (g',u), and multiply both sides by $P_{A(g')-\{u\}}(B)$. Using (A.2.2), (A.2.3), (A.2.4) and the relation (7.1.4) linking x_g^0 and x_g, we obtain

$$x_{g'}^0\ \frac{\lambda(u,g')}{\Lambda(g')}\ P_{A(g')-\{u\}}(B) = \sum_{g\in\bar{G}_u} x_g^0 \sum_{s\in A(g)} p(g,s,g')\ \frac{\lambda(s,g)}{\Lambda(g)}\ P_{A(g)-\{s\}}(B)$$

$$+ \sum_{g\in G_u} x_g^0\ p(g,u,g')\ \frac{\lambda(u,g)}{\Lambda(g)}\ P_{A(g)-\{u\}}(B). \tag{A.2.5}$$

Similarly, multiplying both sides of (7.5.2) by $\hat{P}_{A(g')}(B)$, and using (A.2.2), (A.2.3), (A.2.4), we get

$$x_{g'}^0 \sum_{s\in A(g')\cap\bar{S}'} \frac{\lambda(s,g')}{\Lambda(g')}\ P_{A(g')-\{s\}}(B) =$$

$$\sum_{g\,|\,A(g)\supset A(g')\cap S'} x_g^0 \sum_{s\in A(g)\cap\bar{S}'} \frac{\lambda(s,g)}{\Lambda(g)}\ p(g,s,g')\ P_{A(g)-\{s\}}(B). \tag{A.2.6}$$

The summation of this relation and of relation (A.2.5) when u spans $A(g')\cap S'$ yields (use the birth restriction property)

$$x^0_{g'} \sum_{s\in A(g')} \frac{\lambda(s,g')}{\Lambda(g')} P_{A(g')-\{s\}}(B) = \sum_{g\in G}\sum_{s\in A(g)} x^0_g p(g,s,g') P_{A(g)-\{s\}}(B) \frac{\lambda(s,g)}{\Lambda(g)},$$

that is to say in view of (A.2.1)

$$x^0_{g'} \hat{P}_{A(g')}(B) = P(X(T_1)=g', Y(T_1)\in B), \text{ q.e.d.}$$

Appendix 3. The transition marks (Proof of (8.1.2))

Reproducing computations already done at the begining of the present appendix, we get

$$P^0[X(0)=g, \tilde{\sigma}_0=u_k, X(T_1)=g', Y(T_1)\in B \cap \{x_0(m_{u_k})=0]$$

$$= x^0_g \frac{\lambda(s,g)}{\Lambda(g)} . P_{A(g)-\{u_k\}} . p(g,u_k,g'), \text{ q.e.d.}$$

Appendix 4. Proof of (8.3.5)

By construction,

$$X(0) = X_0, \ \hat{Y}(0) = (\widehat{T_{-T_0}^{X_0} \hat{Y}_0}). \tag{A.4.1}$$

Denote by Z_0 the R.H.S. of (A.4.1). The Palm inversion formula yields

$$P[X(0)=g, \hat{Y}(0)\in\hat{B}] = P[X_0=g, Z_0\in\hat{B}]$$

$$= \frac{1}{E^0_N[T_1]} E^0_N[\int_0^{T_1} 1(X_0 o T_u = g, Z_0 o T_u \in B)\, du]. \tag{A.4.2}$$

But for $u\in[0,T_1)$,

$$X_0 o T_u = X_0$$

and

$$Z_0 o T_u = (\widehat{T_u^{X_0} \hat{Y}_0}).$$

Hence

$$P[X(0)=g,\ \hat{Y}(0)\in B] = \frac{1}{E_N^0[T_1]} E_N^0\left[\int_0^{T_1} 1(X_0=g,\ \widehat{T_u^g\,\hat{Y}_0}\in \hat{B})du\right]. \qquad (A.4.3)$$

In view of the definition of P^0, the R.H.S. of (A.4.3) reduces to

$$x_g^0 \frac{1}{E_N^0[T_1]} E_N^0\left[\int_0^{T_1} 1(\widehat{T_u^g\,\hat{Y}_0}\in \hat{B})du\right] = \frac{x_g^0}{K.\Lambda(g)} P_{A(g)}(\hat{B}),$$

where we have used (A.1.10) and (A.1.14) to get the last expressions. Equation (8.3.5) follows from equation (7.1.4).

Appendix 5. Proof of (9.1.3)

Let

$$d(g) = \sum_{g'\notin G_{s_0}} x_{g'} \sum_{s\in A(g')} p(g',s,g)\,\lambda(s,g') \qquad (A.5.1)$$

and for $g,g'\in G_{s_0}$, let

$$A(g',g) = \delta(g',g)\,(\lambda(s_0,g)-\Lambda(g)) + \sum_{\substack{s\in A(g')\\ s\neq s_0}} p(g',s,g)\,\lambda(s,g'), \qquad (A.5.2)$$

where δ denotes the Kronecker symbol. With these notations, for $g\in G_{s_0}$, the global balance equation of (Σ,Λ) reads

$$x_g\,\lambda(s_0,g) = d(g) + \sum_{g'\in G_{s_0}} x_{g'}\,p(g',s_0,g)\,(\lambda s_0,g') + \sum_{g'\in G_{s_0}} x_{g'}\,A(g',g). \qquad (A.5.3)$$

If $p(.,s_0,.) : G_{s_0}\times G_{s_0}\to \mathbb{R}^+$ is an arbitrary substochastic matrix, the state space G_{s_0} can be partitioned into one inessential class K_0 and a family of essential classes of irreducibility $K_1,K_2,\ldots$ Clearly, $a(g)=0$ (see (9.3.5)) for $g\in K_0$. Let K^+ be the subset of G_{s_0} where $a(g)>0$ and K^- be the subset where $a(g)<0$. Notice that transitions between K^+ and K^- are impossible. In view of equation (9.3.4) the substraction $(\lambda_1.(9.2.8) - \lambda_{s_0}(A.5.3))$ yields the relation

$$- \sum_{g' \in G_{s_0}} b(g')\ A(g',g) + \frac{\lambda_1\lambda_2}{\lambda_{s_0}}\ a(g) =$$

$$(\lambda_1 \Pi - \lambda_{s_0})(d(g) + \lambda_1 \sum_{g' \in G_{s_0}} \hat{x}_{\sigma_1(g')}(p(g',s_0,g) - \delta(g,g'))\ \lambda(s_0,g')), \qquad \text{(A.5.4)}$$

where $b(g) = \lambda_1 \hat{x}_{\sigma_1(g)} - \lambda_{s_0} x_g$, $g \in G_{s_0}$. Let

$J^+ = K^+ \cup \{g \in G_{s_0} \ / \ b(g) > 0\}$ and

$J^- = K^- \cup \{g \in G_{s_0} \ / \ b(g) < 0\}$.

Let $g \in J^+$. Since transitions from K^- to J^+ are impossible, (A.6.4) reads

$$- \sum_{g' \in J^+} b(g')\ A(g',g) + \frac{\lambda_1\lambda_2}{\lambda_{s_0}}\ a(g) = \sum_{g' \in J^-} b(g')\ A(g',g)$$

$$+ (\lambda_1 \Pi - \lambda_{s_0})\ \{d(g) + \lambda_1 \sum_{g' \in K^0} \hat{x}_{\sigma_1(g')}\ p(g',s_0,g)\ \lambda(s_0,g')$$

$$+ \lambda_1 \sum_{g' \in K^+} \hat{x}_{\sigma_1(g')}\ (p(g',s_0,g) - \delta(g,g'))\ \lambda(s_0,g')\}. \qquad \text{(A.5.5)}$$

Assume that $\lambda_1 \Pi - \lambda_{s_0} < 0$, which is always possible. Summation of the right hand side of equation (A.5.5) for all $g \in J^+$ yields a non-positive result (use the relation $\sum_{g \in K^+} p(g',s_0,g) = 1$). Hence the summation of the left hand sides has to yield a non-positive result too, that is

$$\frac{\lambda_1\lambda_2}{\lambda_0} \sum_{g \in K^+} a(g) + \sum_{g' \in J^+} b(g')\ (- \sum_{g \in J^+} A(g',g)) \leq 0. \qquad \text{(A.5.6)}$$

From the very definition, $A(g,g') \geq 0$, $g \neq g'$ and $\sum_{g \in G_{s_0}} A(g',g) = 0$, so that $- \sum_{g \in J^+} A(g',g) \geq 0$.

Hence, the left hand side of (A.5.6) is zero which implies that $K^+ = \emptyset$.

Let us show that K^- is empty too. The substraction ((9.2.8)-Π.(A.5.3) yields

$$\frac{\lambda_2}{\lambda_{s_0}}\ a(g) = \sum_{g' \in G_{s_0}} (\hat{x}_{\sigma_1(g')} - \Pi x_{g'})\ A(g',g). \qquad \text{(A.5.7)}$$

Summation of (A.5.7) entails

$$\sum_{g \in G_{s_0}} a(g) = 0 \tag{A.5.8}$$

(use the convergence of the series $\sum_{g \in G_{s_0}} x_g \, | A(g,g) |$). Hence $K^+ = \emptyset$ entails $K^- = \emptyset$, which concludes the proof of (9.3.1).

APPENDIX 6. Proof of the converse theorem in the general case

Let $C = \{g \in G_{s_0} \, / \, c(s_0,g) = 0\}$. For $g \in C$, the immediate activity assumption entails that both $d(g)$ and the second term in the right hand side of (A.5.3) vanish. Hence the insensitivity balance equation is trivially satisfied for all g in C. Furthermore, if $g \in C$, equation (A.5.4) together with the relation $a(g) = 0$ entail that

$$\sum_{g' \in C} b(g') \, A(g',g) = 0, \quad g \in C. \tag{A.6.1}$$

We can rewrite (A.6.1) as

$$b(g) \, M(g) = \sum_{g' \in C} b(g') \, u(g',g), \quad g \in C, \tag{A.6.2}$$

where $(M(g), u(g,.), g \in C)$ is the substochastic infinitesimal generator on C defined by

$$u(g',g) = \sum_{\substack{s \in A(g') \\ s \neq s_0}} p(g',s,g) \, \lambda(s,g') \tag{A.6.3}$$

and

$$M(g) = \Lambda(g) - u(g,g). \tag{A.6.4}$$

Hence, $b(g) = 0$, at least for all g such that this infinitesimal generator on C is strictly substochastic or equivalently such that

$$\sum_{g' \in C} A(g,g') < 0.$$

Let $g \in C$. Using this fact in equations (9.2.8) and (9.2.9) together with the property $a(g) = 0$, $g \in G_{s_0}$, we get respectively (A.6.5) and (A.6.6) below.

$$- \frac{\lambda_{s_0}}{\lambda_1} \sum_{g' \in G_{s_0}} x_{g'} A(g',g) + \lambda_{s_0} x_g c(s_0,g) = \lambda_2 (1-\lambda_{s_0}/\lambda_1) x_g c(s_0,g)$$

$$+ \Pi (d(g) + \sum_{g' \in G_{s_0}} x_{g'} p(g',s_0,g) \lambda_{s_0} c(s_0,g')) \tag{A.6.5}$$

and

$$-(1 - \frac{\lambda_{s_0}}{\lambda_1}) \sum_{g' \in G_{s_0}} x_{g'} A(g',g) + (1 - \frac{\lambda_{s_0}}{\lambda_1}) \lambda_2 c(s_0,g)$$

$$= (1-\Pi) (d(g) + \sum_{g' \in G_{s_0}} x_{g'} p(g',s_0,g) \lambda_{s_0} c(s_0,g')),$$

where we have used the property "$A(g',g) > 0$, $g \notin C$ implies $\sum_{g \in C} A(g',g) < 0$". Elimination of the term $\sum_{g' \in G_{s_0}} x_{g'} A(g',g)$ yields the desired result. The proof of the converse theorem is thus concluded.

REFERENCES FOR PART II : STATIONARY QUEUEING SYSTEMS

The historical article concerning the stability of stationary queueing systems is :

[11] Loynes R.M. (1962) :
The stability of queues with non independent inter-arrival and service times, Proc. Cambridge Ph. Soc. 58, 497-520.

The proof given in §1 is due to :

[12] Neveu J. (1983) :
Construction de files d'attente stationnaires, Lect. Notes Control and Information Sciences 60. Springer-Verlag, 31-41.

One must mention with respect to this proof

[13] Garsia A. (1965) :
A simple proof of Eberhard Hopf's maximal ergodic theorem, J. Math. and Mech. 14, 381-382.

In this article, Garsia used the sequence $(M_n, n \in \mathbb{N})$ and derived formula (1.2.2) : $E^0[M_\infty \wedge (\tau-\sigma)] \le 0$ which is just Hopf's maximal lemma. Indeed, since $M_\infty \ge 0$

$$E^0[M_\infty \wedge(\tau-\sigma)] \ge \int_{\{M_\infty > \tau-\sigma\}} (\tau-\sigma)\, dP^0 = \int_{\{M_\infty \circ\theta > 0\}} (\tau-\sigma)\, dP^0.$$

In addition,

$$\{M_\infty \circ\theta > 0\} = \bigcup_{k\ge 0} \{\sum_{i=0}^{k} \tau\circ\theta^{-i} < \sum_{i=0}^{k} \sigma\circ\theta^{-i}\}.$$

Therefore (Hopf's lemmma) :

$$\int_{\bigcup_{k\ge 0} \{\sum_{i=0}^{k} \tau\circ\theta^{-i} < \sum_{i=0}^{k} \sigma\circ\theta^{-i}\}} (\tau-\sigma)\, dP^0 \le 0.$$

Queueing theory was implicit in Garsia's article. Neveu fully exploited the queueing idea and provided a very elegant proof of the result :

$$\lim_{n\to\infty} \frac{\sum_{i=0}^{k} \sigma\circ\theta^{-i}}{\sum_{i=0}^{n} \tau\circ\theta^{-i}} = \frac{E^0[\sigma]}{E^0[\tau]} \quad P^0 \text{ - a.s.,}$$

where (P^0,θ) is ergodic and both σ and τ are in $L^1(P^0)$ with $E^0\tau > 0$. Neveu's proof goes as follows : According to the definition of M_n

$$\sum_{i=1}^{n} \sigma\circ\theta^{-i} \leq \sum_{i=1}^{n} \tau\circ\theta^{-i} + M_n.$$

We know that if $E^0[\sigma] > E^0[\tau]$, $M_n \uparrow M_\infty < \infty$ P^0 - a.s. Take $\sigma = \frac{1}{2} E^0[\tau]$. We see from this inequality that if $E^0[\tau] > 0$,

$$\lim_{n\to\infty} \sum_{i=0}^{n} \tau\circ\theta^{-i} = \infty, \quad P^0 \text{ - a.s.,}$$

so that

$$\overline{\lim_{n\to\infty}} \frac{\sum_{i=0}^{n} \sigma\circ\theta^{-i}}{\sum_{i=0}^{n} \tau\circ\theta^{-i}} \leq \inf \{\tfrac{1}{a} \,/\, a E^0[\sigma] < E^0[\tau]\}$$
$$= E^0[\sigma] \,/\, E^0[\tau].$$

Interchanging the roles of σ and τ, we obtain similarly

$$\overline{\lim_{n\to\infty}} \frac{\sum_{i=0}^{k} \sigma\circ\theta^{-i}}{\sum_{i=0}^{n} \tau\circ\theta^{-i}} \geq \frac{E^0[\sigma]}{E^0[\tau]}.$$

Hence the result.

The proof of the existence of stationary states for $G/G/s/\infty$ and $G/G/1/0$ queues (§ 3 and § 4 respectively) are taken from Neveu (already cited) and Flipo respectively :

[14] Flipo D. (1983) :

Steady state of loss systems, Compte rendus de l'Académie des Sciences, tome 297, n° 6.

The latter author also provides sufficient conditions for uniqueness of the workload.

The first construction of the steady state loss system in the general case is due to

[15] Lisek B. (1982) :
A method for solving a class of recursive stochastic equation, Z.f. Wahrsch. 60, 151-162.

The Palm approach to queueing formulae such as Little's formula in § 2 for instance is that of the monograph

[16] Franken P., König D., Arndt U. and Schmidt V., (1981) :
Queues and Point processes, Akademie Verlag, Berlin.

For historical comments on the Palm approach, the reader should consult the above book which contains many detailed examples and the monograph of

[17] Khincin A., (1960)
Mathematical methods in the theory of queueing, Griffin, London (English edition).

The historical articles for queueing formulae are

[18] Little J., (1961) :
A proof of the queueing formula $L = \lambda W$, Operations Research, 383-387.

[19] Takacs L., (1963) :
The limiting distribution of the virtual waiting time and the queue size for a single server with recurrent input and general service times, Sankhya A25, 91-100.

The analysis of the $G/G/1/\infty$ queue in random environment of § 5.2 is due to

[20] Baccelli F., Makowski A., (1986)
Queues in random environment, stability and bounds, Stochastic Models (1986), Vol. 2, n° 2, Marcel Dekker. New York.

and that of the SPRT rule can be found in

[21] Flipo D., (1981) :
Comparaison des disciplines de service des files d'attente G/G/1, Annales de l'IHP, vol. XVII n° 2, pp.191-212.

who gives many more interesting details (and also historical references which we omit here).

The insensitivity problem has a long history recorded in the monograph of Franken, König, Arndt and Schmidt (already cited). The presentation given here is that of

[22] Jansen, U., König, D., Nawrotzki, K. (1979),
A criteria of insensitivity for a class of queueing systems with random marked point processes, Math. Operationsforschung Stat., 3, 379-403.

where the finite case is treated.

The new feature of the present treatment to be noted is the treatment of the infinite case (cardS=∞) in the context of Palm measures.

In the finite case, and for mixed Erlang-type renewal inputs, the insensitivity results have been obtained by

[23] Schassberger, R., (1977)
Insensitivity of steady state distributions of generalized semi-Markov processes I, Ann. Proba. 5, 87-99.

[24] Schassberger, R., (1978)
Insensitivity of steady state distributions of generalized semi-Markov processes II, Ann. Proba. 6, 85-93.

[25] Schassberger, R., (1978)
Insensitivity of steady state distributions of generalized semi-Markov processes with speeds, Adv. Appl. Proba. 10, 836-851.

The Jansen-König-Nawrotzki approach avoids the complicated arguments of weak convergence for the passage from Erlang-type-renewal inputs to renewal and general inputs. See for instance :

[26] Barbour, A.D., (1976)
Networks of queues and the method of stages, Adv. Appl. Proba. 8, 584-591.

[27] Barbour, A.D., (1982)
Generalized semi-Markov schemes and open queueing network. J. Appl. Prob. n° 19, 569-474.

The culmination of this weak-convergence approach is

[28] Helme, F., Schassberger, R. (1979),
Insensitive generalized semi-Markov schemes with point process input, Preprint 502, Technische Hochschule Darmstadt, Fachbereich Mathematik.

where stationary (not merely renewal) inputs are considered as well as an infinite number of sites.

The main result of § 8 is due to R. Schassberger [25]. This author has also established the following weak form of the converse theorem : If S' is a set of insensitive sites and if the stationary distribution of Z(t) is given by the Matthes product form (8.3.5) then, the insensitivity balance equations are satisfied by all the sites of S' : see Jansen, König and Nawrotzki cited above and R. Schassberger [23].

The extention to the infinite case allows a complete proof of the fact that M/G/∞/FIFO is not insensitive.

The main result of § 10 (Poissonian outputs) is due to:

[29] Jansen U. and König D., (1980) :
Insensitivity and steady state probabilities in product form queueing networks, Elektr. Informationsverarb. Kybernetik 16, 385-397.

See also

[30] Melamed B, (1979) :
On Poisson traffic processes in discrete space markovian systems with applications to queueing theory, Adv. Appl. Proba. 11, 218-239.

who gives necessary and sufficient conditions for the state at time t of a stationary Markov chain $(X(t), t \in \mathbb{R})$ to be independent from the past at time t of the stream of privileged transitions. In the situation examined by Melamed, this independence implies that the stream of privileged transitions is Poisson.

BIBLIOGRAPHY

POINT PROCESSES, PALM THEORY

GEMAN D., HOROWITZ J. (1973),

Remarks on Palm measures, Ann. Inst. H. Poincaré 9, .215-232.

GRANDELL, J. (1976),

Doubly stochastic Poisson processes, Lecture Notes in Math. 529, Springer-Verlag, Heidelberg.

JAGERS, P. (1973),

On Palm probabilities, Z. fur W. 26, .17-32.

KALLENBERG, O. (1976),

Random measures, Academic Press.

PAPANGELOU, F. (1974),

On the Palm probabilities of processes of points and processes of lines, in STOCHASTIC GEOMETRY,Ed. Harding and Kendall, J. Wiley and sons, 114-147.

ROLSKI, T. (1981),

Stationary random processes associated with point processes,Lecture Notes in Statistics 5, Springer-Verlag, New York.

CONSTRUCTION OF STATIONARY QUEUES

BACCELLI F., COURCOUBETIS C., REIMAN M. (1986),

Construction of the stationary regime of queues with locking, ATT Bell Labs Tech. Memorandum.

BOROVKOV A.A. (1972),

Continuity theorems for multi-server loss systems, Teor. Ver. Prim. 17, 458-468 (in Russian).

BOROVKOV A.A. (1978),

Ergodic and statibility theorems for a class of stochastic equations and their applications, Teor. Ver. Prim. 23, 241-262 (in Russian).

BOROVKOV A.A. (1980),

Asymptotic Methods in Queueing Theory (in Russian), Nauka, Moscow.

BRANDT A. (1985),

On stationary waiting times and limiting behaviour of queues with many servers. I. The general G/G/m case. EIK. 21 , 1/2, 47-64

BRANDT A. (1985),

On stationary waiting times and limiting behaviour of queues with many servers. II. The G/GI/m/ case. EIK. 21 , 3, 151-162 .

BRANDT A., FRANKEN P., LISEK B. (1984

Ergodicity and steady state existence. Contiuity of stationary distributions of queueing characteristics. - In : Proc. Intern. Seminar on Modelling and Performance Evaluation Methodology, INRIA, Paris - Springer-Verlag, 1984, Lect. Notes in Control and Information Sciences, 60, 275-296.

BRANDT A., LISEK B. (1983),

Recursive stochastic equations and their application to queueing theory. Berlin, Ak. der Wissenschaften der DDR, Inst.für Mathematik, Wiss. Sitzungen zur Stochastik WSS-04/83, 1983.

FRANKEN P. and KALAHNE U. (1978),

Existence, uniqueness and continuity of stationary distributions for queueing systems without delay. Preprint 2/76, Sektion Mathematik, Humboldt. Univ., Berlin, Math. Nachr. 86, 97-115.

LISEK B. (1979),

Construction of stationary state distributions for loss systems, MOS, Ser. Statistics 10, 561-581.

LISEK B. (1981),

Stability theorems for queueing systems without delay, EIK 17, 259-278.

LISEK B. (1982),

A method for solving a class of recursive stochastic equations, Z. fur W.. 60, 151-161

LISEK B. (1985),

Stochastic equations $I_{n+1} = f(X_n, U_n)$. Properties of solutions and applications. Preprint n°103, sektion Mathematik Humbolt Univ., Berlin.

NAWROTZKI K. (1978),

Einige Bemerkingen zur Verwendung der Palmschen Verteilung in der Bedienungstheorie, MOS, Ser. Optimization 9, 241-253.

ROBERT P. (1983),

Thèse de troisième cycle, Laboratoire de Probabilités, Université Paris VI.

WIRTH K.-D. (1982),

On stationary queues with batch arrivals. . EIK 18 ,10/11, 603-619.

INSENSITIVITY AND POISSON OUTPUTS

BARBOUR A.D. and SCHASSBERGER R. (1981),

Insensitive average residence times in generalized semi-Markov processes. Adv. Appl. Prob. 13, 720-735.

BASKETT F. CHANDY M. MUNTZ R. and PALACIOS J. (1975),

Open, closed and mixed networks of queues with different classes of customers. J. Assoc. Comput. Mach. 22, 248-260.

BREMAUD P. (1978a),

On the output theorem of queueing theory, via filtering. J. Appl. Prob. 15, 397-406.

BREMAUD P. (1978b),

Streams of a M/M/1 feedsback queue. Z. fur W. 45, 21-33.

BRUMELLES S. (1978),

A generalisation of Erlang's loss sytem to state dependent arrival and service rates. Math. Operat. Res. 3, 10-16.

BURMAN D.Y. (1981),

Insensitivity in queueing systems. Adv. Appl. Prob. 13, 846-859.

CHANDY K.M., HOWARD J.H., and TOWSLEY D.F. (1979),

Product form and local balance in queueing networks. J. Assoc. Comput. Mach. 24, 25,0-263.

COHEN J.W. (1957),

The generalised Engset formulae. Philips Telecommunication Rev. 18, 158-170.

ERLANDER S. (1967),

A note on telephone traffic with losses. J. Appl. Prob. 4, 406-408.

ERLANG A.K. (1917),

Solution of some problems in the theory of probabilities of significance in automatic telephone exchanges. Post Office Electrical Engineers Journal 10, 189-197.

HELM E. and SCHASSBERGER R.(1979),

Insensitive generalized semi-Markov schemes with point process input .Preprint 502, Technische Hogschule DARMSTADT, Fachbereich Mathematik.

HENDERSON W. (1972),

Alternative approaches to the analysis of the M/G/1 and G/M/1 queues. J. Operat. Res. Soc. Japan 15, 92-101.

HENDERSON W. (1983a),

Non-standard insensitivity. J. Appl. Prob. 20, 288-296.

HENDERSON W. (1983b),

Insensitivity and reversed Markov-processes. Adv. Appl. Prob. 15, 752-768.

HORDIJK A. and VAN DIJK N. (1983a),

Networks of queues. Internation Seminar on Modelling and Performance Evaluation Methodology. INRIA 1, 79-135.

HORDIJK A and VAN DIJK N. (1983b),

Adjoint processes, job local balance and insensitivity for stochastic networks. Bull. Internat. Statist. Inst. 50, 776-788.

JANSEN U. and KOENIG D. (1980),

Insensitivity and steady-state probabilities in product form queueing networks. Elektron. Informationsverarbeit. Kybernet. 16, 385-397.

JANSEN U. and KOENIG D. (1979),

A criterion of insensitivity for a class of queueing systems with random marked point processes. Math. Operationsforsch. Statist., Ser. Optimization, Vol. 10 n°3, 379-403.

KELLY F.P. (1976),

Networks of queues. Adv. Appl. Prob. 8, 416-432.

KELLY F.P. (1979),

Reversibility and Stochastic Networks. Wiley, London.

KELLY F.P. (1982),

Networks of quasi-reversible nodes. In Applied Probability-Computer Science, The Interface : Proceedings of the ORSA-TIMS Boca Raton Symposium, ed. R. Disney, Birkhauser Boston, Cambridge, Ma.

KLEIN M. (1982),

Théorème d'insensibilité. Actes du congrès AFCET "Les mathématiques de l'informatique", Paris pp. 243-253.

KOENIG D. and JANSEN U. (1974),

Stochastic processes and properties of invariance for queueing systems with speeds and temporary interruptions. Trans 7th Prague Conf. Information Theory, 335-343.

KOENIG D. and JANSEN U. (1976),

Eine Invarianzeigenschaft zufalliger Bedienungsprozesse mit positiven Geschwindigkeiten. Math. Nachr. 70, 321-364.

KOENIG D. MATTHES K. and NAWROTZKI K (1967),

Verallgemeinerungen der Erlangschen und Engsetschen Forme ln (Eine Methode der Bedienungstheorie). Akademie-Verlag, Berlin.

KOSTEN L. (1948),

On the validity of the Erlang and Engset loss formulae. Het PTT Bedriff 2, 22-45.

KOVALENKO I.N (1962),

Sur la condition pour que, en régime stationnaire, la distribution soit indépendante des lois des durées de conversation. Ann. Telecom. 17. 190-191.

MATTHES K. (1962),

Zur Theorie der Bedienungprozesse. Trans 3rd Prague Conf Information Theory.

OAKES D. (1976),

Random overlapping intervals : a generalization of Erlang's loss formula. Ann. Prob. 4, 940-946.

SCHASSBERGER R. (1978c),

The insensitivity of stationary probabilities in networks of queues. Adv. Appl. Prob. 10, 906-912.

SEVASTYANOV B.A (1957),

An ergodic theorem for Markov processes and its application to telephone system with refusals. Theory Prob. Appl. 2, 104-112.

STOYAN D. (1978),

Queueing networks-insensitivity and a heuristic approximation. Elektron. Informationsverarbeit. Kybernetik 14, 135-143.

VAULOT, E (1927),

Extension des formules d'Erlang au cas où les durées des conversations suivent une loi quelconque. Rev. Gen. d'Elect. 22, 1164-1171.

WHITT W (1972),

Embedded renewal processes in the GI:GI: queue J. Appl. Prob. 9, 650-658.

WHITT (1974),

Continuity of queues. Adv. Appl. Prob. 6, 175-183.

WHITT (1976),

Continuity of generalized semi-Markov processes. Math. J. Operations Res.

TAKACS L. (1969),

On Erlang's formula. Ann. Math. Statist. 40, 71-78.

WHITTLE P. (1967),

Nonlinear migration processes. Bull. Inst. Int. Statist. 42, 642-647.

WHITTLE P. (1968),

Equilibrium distributions for an open migration process. J. Appl. Prob. 5, 567-571.

WHITTLE P. (1985),

Partial Balance and Insensitivity. J. Appl. Prob. 22, 168-176.

WOLFF R.W. and WRIGHTSON C.W. (1976),

An extension of Erlang's loss formula. J. Appl. Prob. 13, 628-632.

Lecture Notes in Statistics

Vol. 26: Robust and Nonlinear Time Series Analysis. Proceedings, 1983. Edited by J. Fr
W. Härdle and D. Martin. IX, 286 pages. 1984.

Vol. 27: A. Janssen, H. Milbrodt, H. Strasser, Infinitely Divisible Statistical Experiments. V
pages. 1985.

Vol. 28: S. Amari, Differential-Geometrical Methods in Statistics. V, 290 pages. 1985.

Vol. 29: Statistics in Ornithology. Edited by B.J.T. Morgan and P.M. North. XXV, 418 pages.

Vol. 30: J. Grandell, Stochastic Models of Air Pollutant Concentration. V, 110 pages. 1985.

Vol. 31: J. Pfanzagl, Asymptotic Expansions for General Statistical Models. VII, 505 pages. 1

Vol. 32: Generalized Linear Models. Proceedings, 1985. Edited by R. Gilchrist, B. Franci
J. Whittaker. VI, 178 pages. 1985.

Vol. 33: M. Csörgő, S. Csörgő, L. Horváth, An Asymptotic Theory for Empirical Reliabilit
Concentration Processes. V, 171 pages. 1986.

Vol. 34: D.E. Critchlow, Metric Methods for Analyzing Partially Ranked Data. X, 216 pages. 1985.

Vol. 35: Linear Statistical Inference. Proceedings, 1984. Edited by T. Caliński and W. Klonecki. V
pages. 1985.

Vol. 36: B. Matérn, Spatial Variation. Second Edition. 151 pages. 1986.

Vol. 37: Advances in Order Restricted Statistical Inference. Proceedings, 1985. Edite
R. Dykstra, T. Robertson and F.T. Wright. VIII, 295 pages. 1986.

Vol. 38: Survey Research Designs: Towards a Better Understanding of Their Costs and Ber
Edited by R.W. Pearson and R.F. Boruch. V, 129 pages. 1986.

Vol. 39: J.D. Malley, Optimal Unbiased Estimation of Variance Components. IX, 146 pages.

Vol. 40: H.R. Lerche, Boundary Crossing of Brownian Motion. V, 142 pages. 1986.

Vol. 41: F. Baccelli, P. Brémaud, Palm Probabilities and Stationary Queues. VII, 106 pages.

GPSR Compliance

The European Union's (EU) General Product Safety Regulation (GPSR) is a set of rules that requires consumer products to be safe and our obligations to ensure this.

If you have any concerns about our products, you can contact us on ProductSafety@springernature.com

In case Publisher is established outside the EU, the EU authorized representative is:

Springer Nature Customer Service Center GmbH
Europaplatz 3
69115 Heidelberg, Germany

Batch number: 10704763

Printed by Printforce, the Netherlands